STRATEGIC DATA WAREHOUSING

Achieving Alignment with Business

OTHER AUERBACH PUBLICATIONS

Advances in Semantic Media Adaptation and Personalization, Volume 2
Marios Angelides
ISBN: 978-1-4200-7664-6

Architecting Secure Software Systems
Manish Chaitanya and Asoke Talukder
ISBN: 978-1-4200-8784-0

Architecting Software Intensive Systems: A Practitioners Guide
Anthony Lattanze
ISBN: 978-1-4200-4569-7

Business Resumption Planning, Second Edition
Leo Wrobel
ISBN: 978-0-8493-1459-9

Converging NGN Wireline and Mobile 3G Networks with IMS: Converging NGN and 3G Mobile
Rebecca Copeland
ISBN: 978-0-8493-9250-4

Delivering Successful Projects with TSP^{SM} and Six Sigma: A Practical Guide to Implementing Team Software Process^{SM}
Mukesh Jain
ISBN: 978-1-4200-6143-7

Designing Complex Systems: Foundations of Design in the Functional Domain
Erik Aslaksen
ISBN: 978-1-4200-8753-6

The Effective CIO: How to Achieve Outstanding Success through Strategic Alignment, Financial Management, and IT Governance
Eric Brown and William Yarberry, Jr.
ISBN: 978-1-4200-6460-5

Enterprise Systems Backup and Recovery: A Corporate Insurance Policy
Preston Guise
ISBN: 978-1-4200-7639-4

Essential Software Testing: A Use-Case Approach
Greg Fournier
ISBN: 978-1-4200-8981-3

The Green and Virtual Data Center
Greg Schulz
ISBN: 978-1-4200-8666-9

How to Complete a Risk Assessment in 5 Days or Less
Thomas Peltier
ISBN: 978-1-4200-6275-5

HOWTO Secure and Audit Oracle 10g and 11g
Ron Ben-Natan
ISBN: 978-1-4200-8412-2

Information Security Management Metrics: A Definitive Guide to Effective Security Monitoring and Measurement
W. Krag Brotby
ISBN: 978-1-4200-5285-5

Information Technology Control and Audit, Third Edition
Sandra Senft and Frederick Gallegos
ISBN: 978-1-4200-6550-3

Introduction to Communications Technologies: A Guide for Non-Engineers, Second Edition
Stephan Jones, Ron Kovac, and Frank M. Groom
ISBN: 978-1-4200-4684-7

IT Auditing and Sarbanes-Oxley Compliance: Key Strategies for Business Improvement
Dimitris Chorafas
ISBN: 978-1-4200-8617-1

The Method Framework for Engineering System Architectures
Peter Capell, DeWitt T. Latimer IV, Charles Hammons, Donald Firesmith, Tom Merendino, and Dietrich Falkenthal
ISBN: 978-1-4200-8575-4

Network Design for IP Convergence
Yezid Donoso
ISBN: 978-1-4200-6750-7

Profiling Hackers: The Science of Criminal Profiling as Applied to the World of Hacking
Raoul Chiesa, Stefania Ducci, and Silvio Ciappi
ISBN: 978-1-4200-8693-5

Project Management Recipes for Success
Guy L. De Furia
ISBN: 9781420078244

Requirements Engineering for Software and Systems
Phillip A. Laplante
ISBN: 978-1-4200-6467-4

Security in an IPv6 Environment
Jake Kouns and Daniel Minoli
ISBN: 978-1-4200-9229-5

Security Software Development: Assessing and Managing Security Risks
Douglas Ashbaugh
ISBN: 978-1-4200-6380-6

Software Testing and Continuous Quality Improvement, Third Edition
William Lewis
ISBN: 978-1-4200-8073-5

VMware Certified Professional Test Prep
John Ilgenfritz and Merle Ilgenfritz
ISBN: 9781420065992

AUERBACH PUBLICATIONS

www.auerbach-publications.com
To Order Call: 1-800-272-7737 • Fax: 1-800-374-3401
E-mail: orders@crcpress.com

STRATEGIC DATA WAREHOUSING

Achieving Alignment with Business

Neera Bhansali

CRC Press
Taylor & Francis Group
Boca Raton London New York

CRC Press is an imprint of the
Taylor & Francis Group, an **informa** business
AN AUERBACH BOOK

CRC Press
Taylor & Francis Group
6000 Broken Sound Parkway NW, Suite 300
Boca Raton, FL 33487-2742

First issued in paperback 2019

© 2010 by Taylor and Francis Group, LLC
CRC Press is an imprint of Taylor & Francis Group, an Informa business

No claim to original U.S. Government works

ISBN-13: 978-1-4200-8394-1 (hbk)
ISBN-13: 978-0-367-38534-7 (pbk)

Library of Congress Cataloging-in-Publication Data

Bhansali, Neera.
　　Strategic data warehousing : achieving alignment with business / Neera Bhansali.
　　　　p. cm.
　　Includes bibliographical references and index.
　　ISBN 978-1-4200-8394-1 (hardcover : alk. paper)
　　1. Data warehousing. 2. Business--Data processing. I. Title.

　　QA76.9.D37B53 2010
　　651.8--dc22
　　　　　　　　　　　　　　　　　　　　　　　　　　　　　　　　　　　　　2009021246

Visit the Taylor & Francis Web site at
http://www.taylorandfrancis.com

and the CRC Press Web site at
http://www.crcpress.com

Dedication

To Shrenik and Divya

Contents

Preface

If you don't know where you are going, any road will get you there.

– Lewis Carroll

One of the motivations for writing this book was to give today's IT and business managers a perspective of the data warehouse as a corporate asset that can be put to strategic uses. The role of CIOs has changed from just managing the IT infrastructure to being active partners in the management of the business of the company. Because most IT managers possess a technical background and are not skilled in business strategies, this book is written to facilitate their looking at and using the data warehouse from a new perspective. It will help them in aligning the data warehouse to business strategies.

This book is also written for the business managers who provide the resources to build the data warehouses and are often disappointed with the result. It provides them with an understanding of what a data warehouse is and how they could become active partners in leveraging this powerful information resource.

The book focuses on the challenges in aligning data warehouses to business goals and strategies. It provides an understanding of the principles and techniques for strategic alignment and their application to real-world practice of data warehousing.

This book is targeted toward practitioners of data warehouses as well as business executives planning and implementing data warehouses in an organization. Professionals in information systems (IS) and information technology (IT), business management, business administration, and business analysis will find this book useful. Students in a variety of courses including management information systems (MIS), database design, data management, data warehousing, decision support systems, and business administration will find it beneficial to understand how theory is put into practice.

Strategic alignment of the data warehouse to business strategies is a continuous process.

> However beautiful the strategy, you should occasionally look at the results.

— Winston Churchill

Although it was difficult to discuss every aspect of strategic alignment of data warehouses in complete detail in a book of this length, I hope that this book conveys the challenges, potential, and excitement of strategic data warehousing.

Neera Bhansali

Acknowledgments

I wish to thank Mr. Kim Ross, CIO for Nielsen Media Research, Mr. Timothy Eitel, CIO for Raymond James Financial, and Mr. Mark Abbott, Vice President Software Development for Raymond James Financial, for enabling the case studies. I also wish to acknowledge all the participants in the case studies.

I wish to especially thank my family for their love, support, and understanding during the extended time working on this book.

About the Author

Neera Bhansali, PhD, received her doctoral and masters degrees in business from RMIT University, Australia, and BA from Calcutta University, India. Dr. Bhansali is an expert in areas of strategic planning, strategic alignment, data governance, and data warehousing. Over the past twenty years Dr. Bhansali has facilitated transformations and provided strategic direction to organizations in manufacturing, airline, consulting, media, finance, and healthcare industries in Asia, Australia, and North America.

Chapter 1

Introduction

This book explains the role of strategic alignment between business and data warehouse plans in an organization and the role of that alignment in successful adoption of a data warehouse. It addresses the question, What role does strategic alignment play in the successful adoption of the data warehouse? A data warehouse is a collection of data from multiple sources, integrated into a common repository and extended by summary information for the purpose of analysis (Ester et al., 1998). This repository allows enterprises to collect, organize, interpret, and leverage the information (data) they have for decision support (Wixom and Watson, 2001; Gupta and Mumick, 2005; Groth, 2000; Gardner, 1998; Sethi and King, 1994). It provides the foundation for effective business intelligence solutions for companies seeking competitive advantage (Chenoweth et al., 2006).

The use of information technology in business has transformed over the last several decades from operational utility in the 1960s to that of a competitive weapon today (Carr, 2003; Kayworth et al., 2001; Ives and Learmonth, 1984; Bakos and Treacy, 1986). This phenomenon has affected the ways organizations are managed as well as the way IT affects the strategic activities of an organization (Pollalis, 2003). The strategic use of information technology has become a fundamental issue for every business because information technology can enable the achievement of competitive and strategic advantage for the enterprise (Kearns and Lederer, 2000; Luftman et al., 1993; Jarvenpaa and Ives, 1991).

In today's era of globalization (Breathnach, 2000; James, 1999), the prevailing hyper-competitive markets (Eustace, 2003; Gagnon, 1999) bring pressure for businesses to shorten product life cycles (Bussmann, 1998; Griffin, 1997), quickly identify and penetrate new market segments, and increase operational efficiencies (Krishnan et al., 1999; Mooney et al., 1996). Businesses seek sustainable competitive advantage in these markets by leveraging technology to the fullest extent (Alavi and

Leidner, 1999). With strong competition and growing need for information, enterprises are eager to obtain fast and accurate information for better decision making (Dean and Sharfman, 1996). Companies are continuously investing in processes and technologies that enable better, faster, and more accurate decision making (Hurd, 2003). One such enterprise decision-making platform is a data warehouse.

Data Warehouse

"Data warehouse is a subject-oriented, integrated, non-volatile, time-variant collection of data in support of management's decision making process" (Inmon, 1996a). The concept of integrated data for management support is not a new one. Management information systems and executive information systems have been around since the early 1970s (Shim et al., 2002). However, the operational IT environment in most large companies is very heterogeneous as a result of decades of changing technologies (March et al., 2000). Data resides in legacy systems in various technologies and environments, ranging from PCs to mainframes (Robertson, 1997). As a result, they are incapable of supporting management decision processes due to a lack of data integration. Data warehouses offer data integration solutions and improved access to timely, accurate, and consistent data (Ang and Teo, 2000; Ingham, 2000). A data warehouse equips its users with effective decision support tools by integrating corporate-wide data into a single repository from which users can run reports and perform ad hoc data analysis (Meyer and Cannon, 1998). The data warehouse leverages the investments already made in legacy systems, allowing business users the potential for much greater exploitation of informational assets (Counihan et al., 2002). A data warehouse helps reduce the costs, increases value-added activities, and improves efficiency (Zeng et al., 2003a).

The data warehouse provides effective business decision support data to an organization (Poe et al., 1998). Some of the successful companies that have leveraged this data effectively include Wal-Mart (Westerman, 2001), Amazon (Rundensteiner et al., 2000), Citigroup (Altinkemer, 2001), and Nielsen Media Research. The strength of the data warehouse is its organization and delivery of data in support of management's decision-making process (Meyer and Cannon, 1998). The data warehouse supports decision making and business analyses by integrating data from multiple, incompatible systems into a consolidated database (Inmon, 1996).

The data warehouse also allows sophisticated analyses of data. The capability of the data warehouse to perform the analysis has been documented by J. Srivastava and Chen (1999). In the data warehouse, data is periodically replicated from operational databases and external providers of data, and is conditioned, integrated, and transformed into a read-only database to discern patterns of behavior, support decision support systems, and enable online analytical processing. Little and Gibson (2003) state that data warehouses also help in accessing, aggregating, and analyzing large amounts of data from diverse sources to understand historical performance or behavior and to predict and manage outcomes.

Data warehouse technology is inherently complex (Gardner, 1998; Chaudhuri and Dayal, 1997), requires huge capital spending (Wixom and Watson, 2001), and consumes a lot of development time. The complexity of data warehouse implementations is a subject of ongoing studies (L. Chen et al., 2000; H. Lee et al., 2001; Klenz, 2001; Sperley, 1999). The adoption of data warehouse technology is not a simple activity of purchasing the required hardware and software, but rather a complex process to establish a sophisticated and integrated information system (Vassiliadis et al., 2000; Wixom and Watson, 2001). Building a data warehouse consists of a complex process involving data sourcing, data extraction and conversion, population of the data warehouse database, data warehouse administration, creation of metadata, and access to the data warehouse database for decision support and business intelligence (Little and Gibson, 2003; Berndt and Satterfield, 2000; Manning, 1999; Shahzad, 1999; O'Sullivan, 1996; Watson et al., 2004).

Challenges in Data Warehousing

In the past decade there has been an explosive growth in products and services offered for the adoption of data warehouse technologies (Datta et al., 1998; Meyer and Cannon, 1998; Koch, 1999). Data warehousing has also been a rapidly growing area in management information systems (Gary, 2004). Vassiliadis (2000) in his study concludes that the area of data warehousing is thriving and there is potential for further growth, but he adds that data warehouse projects are very risky. Companies are integrating their data and building data warehouses to create the advantages of identifying new markets for products and services, providing improved customer service, and retaining customer loyalty (Berry and Linoff, 1999; Rygielski et al., 2002), and reducing production and inventory costs (M. Fisher and Raman, 1996). But a review of the literature suggests that a majority of data warehouse projects have a high possibility of failure (Chenoweth et. al., 2006; Hwang et al., 2004; Wen et al., 1997; Watson et al., 1999; Vatanasombut and Gray, 1999; Kelly, 1997), and many firms are failing to realize the benefits of data warehousing (Johnson, 2004).

Thus, even though data warehouses have emerged as a powerful tool in delivering information to users, creating competitive advantage (Groth, 2000; Berson et al., 2000; Gardner, 1998), and building support for decision making (Gray and Watson, 1998; Shim et al., 2002) and customer satisfaction (Berry and Linoff, 2004; Xu et al., 2002; Hui and Jha, 2000), their implementation is not always successful. Although the data warehousing concept continues to attract interest, many data warehousing projects are not only failing to deliver the benefits expected of them, as discussed in the previous paragraph, they are proving to be excessively expensive to develop and maintain (Manning, 1999). According to Koch (1999), 50% of these multimillion dollar projects fail to meet the desired levels of success.

The reality of data warehousing is much more risky and difficult than the promise. One of the most recent, high-profile, and highly visible failures of data warehouses was the Virtual Case File (VCF) commissioned by the FBI, costing more than $175 million (Goldstein, 2005). The VCF was commissioned as a response to the September 11, 2001, incident, to allow U.S. federal agents and intelligence agencies to share vital investigative information and develop a system to help spot patterns that might signal a future attack by terrorists on the United States of America. This failure has been the subject of a study by Goldstein (2005). He suggests that the organizational structure, communication, and implementation were the key reasons for failure. Vassiliadis (2000) in his study of data warehouses has identified sociotechnical and procedural factors that contribute to the failure of data warehouses, apart from design and technical factors.

Goal

The objective of this book is to help the reader understand the role of strategic alignment in the success of data warehouse implementation. Although various causes have been attributed, ranging from technical to organizational reasons, for failure of data warehouses, the underlying strategic alignment issues have not been understood in detail. Different factors have been identified that affect successful implementation of data warehouses, which include project sponsorship (Hwang et al., 2004), architecture selection (Zhou et al., 2000; Little and Gibson, 2003; Tyagi, 2003; Peacock, 1998; Inmon, 1998a; Murtaza, 1998; Sigal, 1998; Van Den Hoven 1998), technological sophistication (Triantafillakis et al., 2004; Zeng et al., 2003; J. Srivastava & Chen, 1999; Shahzad, 1999; Sigal 1998), user participation (Gorla, 2003; Guimaraes et al., 2003; Nah et al., 2004), and data quality (Y.W. Lee et al., 2004; Sinn, 2003; Fisher et al., 2003; Armstrong, 1997; Redman, 1995).

However, each data warehouse system has an organizational specific set of requirements, constraints, issues, and implications that need to be addressed. There is no "one strategy fits all" solution to these problems. It can be easily envisioned that a standard approach to all projects is not feasible. Every data warehouse has its own issues of architecture, design, technology, data quality, and users that change with every organization. Addressing these factors alone, as was attempted in the VCF study (Goldstein, 2005), does not guarantee the implementational success of the data warehouse. This book postulates that success depends on being able to align the data warehouse to the business plans and strategy. It explores and answers the question, What role does the alignment of the data warehouse to business plans and strategies have in the success of data warehouse adoption?

The challenge of aligning the data warehouse to the business strategy is at the heart of this book. The achievement of this alignment is important as data warehouses are being built to advance the strategic initiatives of an organization (Raghupathi and Tan, 2002; Cooper et al., 2000; Gray and Watson, 1998). Data warehouse technology enables the strategic use of information (Sammon and Finnegan, 2000). According to Counihan et al. (2002), data warehousing had emerged as a response to the problems encountered by those trying to implement decision support systems for strategic management. With the substantive and persuasive changes that this technology is enabling, it is no longer possible to have a disconnect between an organization's strategic plans, goals, and directions and its IT initiatives, resources, and management (Hitt, 2001).

Strategic Alignment

Organizations allocate considerable resources to data warehouse projects, but there has been very little discussion on how to achieve strategic alignment between the data warehouse and the business plans, to ensure its success. Discussion on managerial or strategic issues of data warehousing have been rare. There is no book that empirically investigates the relationship between strategic alignment and data warehouse success. Although the need for commitment and support from top management has been identified as a critical factor (Wixom and Watson, 2001), no specific guidelines have been discussed on how to attain this. One of the questions that needs to be answered is whether strategic alignment can resolve managerial and strategic issues in data warehousing.

The number of technologies and software capabilities that exist are more than what a business could ever possibly adopt. The key issue for companies is not the availability of technology but choosing which technology to deploy and to what purpose. Businesses have invested billions of dollars in information technology to date, yet studies like those of Ryan and Harrison (2000) indicate that more than 50% of IT implementations actually cost more than twice their original estimates, and the same can be said of data warehouse implementations (Wixom and Watson, 2001). A lack of foresight in the IT investment decision process has been cited for this diminishing payoff (Schniederjans and Hamaker, 2003); others cite a need to deploy information technologies in ways that are of the most relevance to the business and its strategic objectives (Andal-Ancion et al., 2003; Kearns and Lederer, 2003; Tallon et al., 2000). This is applicable to data warehouses, too. Data warehouses are large, expensive projects (Manning, 1999) often built to address the strategic objectives of an organization (Raghupathi and Tan, 2002; Cooper et al., 2000) and have had a high rate of failure (Hwang et al., 2004; Chenoweth et al., 2006) in realizing benefits.

Current Status of Strategic Alignment Research

The concept of strategic alignment is more than two decades old (McLean and Soden, 1977; Chan, 1996; Henderson and Venkatraman, 1990), but improving IS and IT strategic planning continues to rank among the major issues facing IT executives. Strategic alignment has been identified as one of the most critical IS issues (G.G. Lee and Bai, 2003). IS strategic alignment has been among the top five challenges faced by senior executives over the past few decades (Chan, 1996; Chan and Huff, 1993a) and continues to be of increasing importance today. The reason for the interest in strategic IS alignment is because it has been shown to enhance not only IS success but organizational success as well (Hirschheim and Sabherwal, 2001).

Several frameworks have been proposed to study and explore alignment between IT and business strategy (Henderson and Venkatraman, 1990; Chan, 1996; N. Shin, 2001; Tallon et al., 2000; Kearns and Lederer, 2003; Prahalad and Krishnan, 2002; Loebbecke and Wareham, 2003; Pollalis, 2003; Bai and Lee, 2003; Maes, 2000; Burn, 2000; Hirschheim and Saberwal, 2001). The concept of strategic alignment proposed by Henderson and Venkatraman (1990) is based on two building blocks: strategic fit and functional integration. Findings by Chan (1996) lend support to the view that examining isolated components of strategy and performance can be misleading. N. Shin (2001) provides empirical evidence for the importance of aligning IT with business strategies such as vertical disintegration and diversification. Tallon et al. (2000) focused on process-level measures and found that management practices such as strategic alignment and IT investment evaluation contribute to higher perceived levels of business value.

Studies by Kearns and Lederer (2003) provide an explanatory framework of the alignment–performance relationship within the context of a resource-based view and furnish several new constructs. The issue of flexibility in strategic planning has been explored by several researchers (Loebbecke and Wareham, 2003; Prahalad and Krishnan, 2002; Wixom and Watson, 2001). Loebbecke and Wareham (2003) find that strategy and strategic planning needs to embrace greater flexibility to nurture creativity and innovation.

Bai and Lee (2003) investigate the organizational factors that influence the quality of the IT strategic planning process and the organizational mechanisms for success in strategic planning. Burn and Szeto (2000) contend that effective alignment of IT and business strategies can be attained by means of strategic information systems planning (SISP).

Although many researchers, as noted in this text, have explored alignment between IT and business strategy, discussion in this field has largely been confined to theoretical issues and practical generalizations. Notwithstanding the importance of the strategic alignment model, it is difficult to apply the model in practice (Van Eck et al., 2004). Van Eck et al. contend that given a particular alignment case study, there are no objective, concrete criteria to determine which of the alignment perspectives play a role in the case. No study or book has focused on how organizations actually achieve alignment while implementing data warehouses.

The Gap

The gap between practitioners and researchers is widely discussed in the IT community. The situation regarding data warehousing follows the same pattern — practitioners complain that their practical problems are overlooked by research (Vassiliadis, 2000). There is very little reference to alignment issues in data warehousing. Little insight exists to guide the successful development and implementation of data warehouses that align to business goals and strategies. No study has focused on how organizations actually achieve alignment while developing or implementing a data warehouse project. No practical design guidelines exist to achieve strategic alignment at an operational level. There is little empirical evidence on how to carry out alignment in a large project like a data warehouse.

Most authors focus on the technological and operational aspects of data warehouses. Very little discussion addresses the managerial or strategic aspects of a data warehouse. These aspects have a significant impact on data warehouse adoptions but have received very little attention so far. This book attempts to address this gap.

This book examines the impact of strategic alignment on successful adoption of a data warehouse. The goal of this book is to develop an understanding among practitioners, managers, and data warehouse users, to facilitate the alignment between data warehouses and business strategy. Interest in strategic alignment of the data warehouse implementation to the business strategy is warranted because organizations are unable to realize sufficient value from their investments in data warehouses (Wixom and Watson, 2001; Frolick and Lindsey, 2003; Wells and Thomann, 1995). This book postulates that alignment of the data warehouse strategy to business strategy would contribute to greater success of data warehouse implementations.

This book reviews existing practices in data warehousing and strategic alignment and then presents factors that may be useful in implementing data warehouses. It identifies and classifies factors that may facilitate the alignment between business strategy and data warehouse projects. Because data warehouse implementations are expensive and time consuming, concentrating on the strategic alignment factors could result in economic benefits for the organization. A practical model of how to achieve such alignment could result in significantly improved and successful data warehouse implementations. This provides a set of comprehensive factors that can be used to align the data warehouses to business strategy and plans.

A pathway for facilitating successful adoption of a data warehouse is provided in this book. The factors presented in this book allow data warehouse professionals to ensure that their project, when implemented, will achieve the strategic goals and business objectives of the organization. The model presented in this book and the strategic alignment factors identified will guide the data warehouse participants at all levels in uncovering and addressing data warehousing alignment issues that may previously have remained untouched.

A secondary contribution of this book is the development of a set of interview and questionnaire instruments to capture data that impact the alignment of data warehouses to business strategy. The questions have proven to be effective in eliciting responses that facilitate the evaluation of the data warehouse in different industries.

Case Studies

Two cases in two different industry sectors, media (Nielsen Media Research) and finance (Raymond James Financial), are presented in this book. The two case studies address strategic alignment of data warehouses to business goals and objectives. The case studies provide the organizational context for the study of the relationship between data warehouse technology and business strategy. The case studies explore in depth the complexities and processes of a data warehouse and its alignment to business strategies and goals.

Organization of Remaining Chapters

Chapter 2 presents the evolution of information technology and the benefits of data warehousing. Chapter 3 presents in detail the difference between data warehouse and traditional operational systems. Chapter 4 discusses the complexity of the data warehouse development process and its components, including data sourcing, data conversion and extraction, the data warehouse database management system, data warehouse administration, business intelligence tools, and metadata. Chapter 5 presents the data warehouse architectures. Chapter 6 discusses the factors affecting the success of a data warehouse. These include organization factors, user factors, technology factors, and data factors. Chapter 7 discusses strategic alignment with respect to IT. It stresses the need for aligning IT with business strategy, notes recent developments in strategic alignment research, and presents the alternatives to the strategic alignment model. It discusses the enablers of alignment between IT and business. Chapter 8 presents in detail the factors that align data warehouses to business strategies and goals and ways to achieve strategic alignment in data warehouses. Chapter 9 presents a case study at Nielsen Media Research. Chapter 10 presents a case study at Raymond James Financial. Chapter 11 discusses how to assess strategic alignment of data warehouses in organizations. Chapter 12 presents an epilogue.

Chapter 2

Benefits of a Data Warehouse

Evolution of Information Technology

There is a remarkable parallel between the evolution of species and evolution of information technology. Human beings are one of the great success stories of evolution. Human beings far surpass other animals in their ability to process information and adapt to changing environments. Within an ecosystem, a species must adapt to numerous relationships with predators, prey, and competitors. Life without information processing is virtually impossible. In order to survive, an organism has to receive information, process it, and react to the stimuli. Surviving requires complex information processing and adaptations.

Knowledge and innovation have played a crucial role in development from the beginnings of human history. Civilizations have advanced as people discovered new ways of exploiting various physical resources. Humans have used technical means to communicate and to process information from very early on. In the 20th century, for the first time information processing was able to be performed outside the human brain.

The history of information technology has moved from one stage to another through the ages. Communications through speech and paintings in 3000 B.C. progressed through indentations in clay tablets to the use of symbols by Phoenicians in 2000 B.C., to the use of paper by the Chinese in 100 A.D. The invention of the numbering system by the Hindus around 200 A.D. progressed through the abacus to early computers (literally computing numbers) such as slide rules, the Pascaline, and Babbage's engines to analytical engines in the 1800s.

With the invention of electricity and the beginning of telecommunication, knowledge and information was converted to electrical impulses and processed with increasing speed, progressing from valves to transistors and then to integrated circuits and present-day microprocessors on a single chip. Globalization and the technological revolution of the last few decades have made knowledge and information the key driver of competitiveness. Information now plays a central role in economic growth and development. The growth of the digital universe accompanied by an explosion of information now has implications for every aspect of life for an individual and organizations.

Evolution of Information Systems

Data is often considered the building block of information systems. Data is defined as a gathered body of facts. In the context of an organization, data is the factual information used as a basis for reasoning or calculation. Data by itself, without context, is devoid of information (Figure 2.1).

Data must be processed to be meaningful. Information is that which is communicated or received concerning a particular fact or circumstance. Drawing various pieces of data together within an appropriate context yields information that is useful. It is amenable to analysis and interpretation. Knowledge is considered to be the sum of what is known. It is the general understanding and awareness garnered from accumulated information, enabling new contexts to be envisaged (Figure 2.2).

Knowledge is valuable and grows by an evolutionary process. Knowledge increases an individual's and an organization's capacity to take effective action. To gain valuable knowledge more rapidly, organizations build complex information

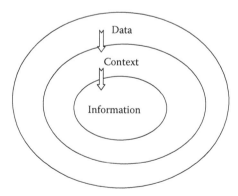

Figure 2.1   Data in context provides information.

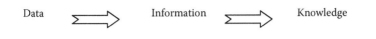

Data ⟹ Information ⟹ Knowledge

Figure 2.2   From data to knowledge.

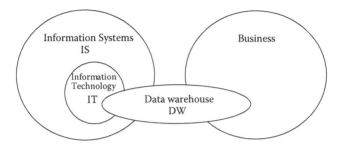

Figure 2.3   Interdependency of terms.

processing systems. The interdependency of the terms used in this chapter is schematically shown in Figure 2.3.

Information systems include systems that perform storage and maintenance of representations of structured information and also presentation of structured output information from the stored representations on request. With the dramatic development of data storage technology, processing speed, and representation size of information, information systems have become increasingly sophisticated. Processing information has moved from sequential files on magnetic tapes to relational databases on mainframes to Web-enabled graphical applications.

Information systems evolved from batch processing of individual files to integrated real-time, online processing systems. These operational systems can maintain and present an up-to-the-second, consistent picture of an enterprise's resources. The up-to-date picture allows the management of a corporation to use its valuable resources much more effectively. Integration of the data helps achieve an up-to-the-second, consistent picture of the business.

These operational systems soon led to the development of analytical systems. Management information systems and executive systems developed that performed more complex analyses of information for long-range planning. These decision support systems (DSS) did not need to be up-to-the-second as operational systems. In fact, information representations stored in operational transaction systems were extracted on a daily or weekly basis to populate the databases supporting analytical functions. DSS supported complex decision making and problem solving.

With the decline in costs of computing power and the advent of larger and more powerful databases, DSS have evolved significantly since their early development in the 1970s. The evolution of information technology infrastructures parallels the three eras of growth in the computer industry — the data processing era, the microcomputer era, and the network era. In the last decade, with the advent of the Internet, World Wide Web, and telecommunications technology, organizational environments have increasingly become more global, complex, and connected. The World Wide Web's impact on decision making has been to make the process more efficient and more widely used. Information technology is being used to improve the *efficiency* with which a user makes a decision, and also to improve the *effectiveness* of that decision.

With the natural evolutionary advancements in information technology and database technologies, data warehouses have emerged as a powerful tool to build decision support systems. Along with the Web, online analytical processing (OLAP), and data mining, the power of data warehouses to support organizational decision making has increased vastly.

Benefits of a Data Warehouse

William Inmon (1996a) defined a data warehouse as "[a] subject-oriented, integrated, non-volatile, time-variant collection of data in support of management's decision-making process." A data warehouse equips its users with effective decision support tools by integrating company-wide data into a single repository from which end users can run reports and perform ad hoc data analysis (Meyer and Cannon, 1998). A data warehouse helps reduce costs, increases value-added activities, and improves efficiency (Zeng et al., 2003a). It is said to provide enhanced data storage and data access functionality.

Organizations are increasingly recognizing the possibilities and implications of data warehousing. Data warehousing has been found to be a useful technology for a large number of modern applications (Rundensteiner et al., 2000). Such applications range over diverse domains such as business, leisure, health (Schubart and Einbinder, 2000), science, libraries, and education. Today, improved access to timely, accurate, and consistent data needs to be shared easily with team members (Little and Gibson, 1999), decision makers, and business partners for efficient decision making. Many companies have recognized the strategic importance of knowledge hidden in their large databases and have therefore built data warehouses (Ester et al., 1998). Data warehouses provide the foundation for effective business intelligence solutions for companies seeking competitive advantage (Chenoweth et al., 2006). The data warehouse provides six main benefits to an organization: (1) decision support, (2) data analysis, (3) data mining, (4) data integration, (5) improved efficiency, and (6) customer management (Figure 2.4).

Decision Support

Initially, information technology projects were undertaken to provide cost savings over manual systems. Turbulent, faster environments led to shorter decision-making

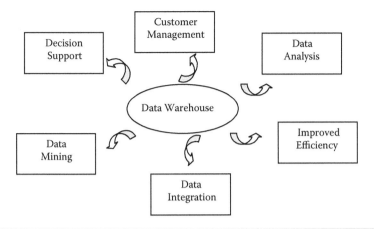

Figure 2.4   Benefits of data warehousing.

cycles. This led to the development of DSS and enterprise information systems (EIS). Later, technological advances, business pressure, competitive pressures, and management interest led to the development of data warehouses. Ultimately, the Management Information Systems/Decision Support System (MIS/DSS) family of systems gradually evolved into data warehousing systems.

The strength of the data warehouse is its organization and delivery of data in support of management's decision-making process (Meyer and Cannon, 1998). Although EIS and DSS were very useful, they lacked a strong database behind them. Information gathered for efficient running of day-to-day business could not be used directly for another purpose. Managerial decision making required consideration of the past and future and not just present transactional information. As a result most DSS builders created their own databases, an area in which they were not the prime experts.

Traditional databases are incapable of handling the demands for increasing online information retrieval, access, update, and maintenance. These limitations affect the managements' efficiency and ability to make reliable decisions in a timely manner (Hwang et al., 2004). A data warehouse is an effective way to provide business decision support data by integrating information and making it available for querying and analysis (Widom, 1995). It is a way of organizing business information that can provide better visibility to management and more insight than the traditional information systems used to support day-to-day operations (Williams, 1999).

Data Analysis

A data warehouse is a repository of integrated information for querying and analysis. Data warehousing is gaining popularity as organizations realize the benefits of being able to perform sophisticated analyses of their data (J. Srivastava and Chen,

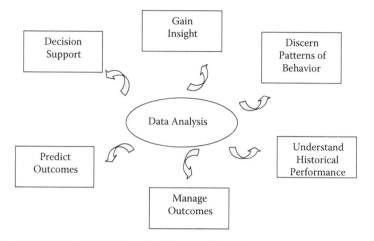

Figure 2.5   Data analysis.

1999). The data warehouse provides opportunities for performing data mining tasks such as classification and clustering (Ester et al., 1998). The advantage of a data warehouse in an organization is that it allows decision makers to analyze data without interfering with the transaction processing system. In a data warehouse, data is periodically replicated from operational databases and external providers of data, and is conditioned, integrated, and transformed into a read-only database to discern patterns of behavior, support DSS, and enable OLAP (Little and Gibson, 2003). Data warehouses also help in accessing, aggregating, and analyzing large amounts of data from diverse sources to understand historical performance or behavior and to predict and manage outcomes (Figure 2.5).

Large data warehouses lead to increased analysis and use of the accumulated historical DSS data. Data warehouses help managers notice problems and opportunities, increase the extent of their analysis, and lead to better decisions. Historical, summarized, and consolidated data from several operational databases over potentially long periods of time is analyzed in a data warehouse using OLAP tools. OLAP is a category of software technology that enables analysts, managers, and executives to gain insight into data through fast, consistent, interactive access to a wide variety of possible views of information that has been transformed from raw data.

OLAP differs from online transaction processing (OLTP) systems. OLTP applications automate clerical data processing tasks such as order entry or sales transactions that are the basic day-to-day operations of an organization. These tasks are structured and repetitive and consist of short, atomic, isolated transactions. The transactions require detailed, up-to-date data. The database is designed to reflect the operational semantics of known applications. OLTP systems are often built using relational databases.

On the other hand, OLAP tools provide multidimensional analysis of consolidated enterprise data supporting end-user analytical and navigational activities. OLAP operations include rollup (increasing the level of aggregation) and drilldown (increasing detail) along one or more dimension hierarchies. They provide slice-and-dice (selection and projection) and pivot (reorienting the multidimensional view of data). Both relational and multidimensional database technologies are used for OLAP. Relational structures like the star schema are preferred for very large data warehouses.

Data Mining

Data mining is often called database exploration or information and knowledge discovery. Data mining allows sophisticated data search capabilities that use tools to discover patterns and correlations in data. Data mining finds and extracts knowledge buried in corporate data warehouses, which leads to improvements in the understanding and use of the data.

Dramatic increases in networking, storage, and processing technologies have led to the creation of very large databases that record vast amounts of transactional information. The rapidly expanding volume of real-time data, resulting from the explosion in activity from the Web and electronic commerce, has contributed to the demand for data mining tools. In tandem with this increase in data, OLAP tools have become more powerful in recent years. Along with artificial intelligence and statistical tools, more sophisticated data analysis is possible. Data mining tools find patterns in data and infer rules from them. Data mining allows discovery of hidden patterns, predictive modeling, and forensic analysis.

The automated, future-oriented analyses made possible by data mining move beyond the analyses of past events typically provided by history-oriented tools such as DSS. Data mining tools answer business questions that in the past were too time consuming to pursue. For instance, it is now even possible to conduct real-time data mining analysis of "clickstream data" for organizations that have highly interactive websites that generate a lot of data. Data mining allows organizations to grasp the buying patterns of visitors on their websites and enhance customer relationship management.

There is a symbiotic relationship between the activity of data mining and the data warehouse (Inmon, 1996b). A data warehouse contains integrated, summarized, and detailed as well as historical data. A data warehouse also contains metadata. Each of these elements found in the data warehouse enhances the data mining process. Integrated data saves the data miner time that would otherwise be spent on data cleansing and transforming raw data to make it usable for effective data mining. In the data warehouse, the data contained is already reconciled and the data structures are standardized. The data warehouse contains detailed as well as summarized data. Data mining processes can be applied to the granular detailed level of data to uncover patterns or at higher levels of available summarized data to save time. The historical data contained in the data warehouse allows the data miner to compute trends or to discover patterns of behavior, such as studying business cycles.

Enhanced Integrated Data

The integration of enterprise-wide data is frequently a time-consuming and cumbersome activity. Integration of data is necessitated by circumstances such as mergers and acquisitions, or situations arising where data is produced from different software platforms that are incompatible. Integration of data within the organization is required when data is produced by software systems evolving independently of each other for specific purposes. Enterprise-wide integration of data pulls from multiple sources and makes them available for supporting management decisions. In a data warehouse, the data from heterogeneous operational systems that may be incompatible is extracted, cleansed, and integrated to support decision-making processes.

A data warehouse supports decision making and business analyses by integrating data from multiple, incompatible systems into a consolidated database (Meyer and Cannon, 1998). When data is brought into a data warehouse, inconsistencies are removed in both nomenclature and conflicting information. Data is integrated so that it is referred to only in one way, has the same format, and has the same units for measuring attributes. Data in the data warehouse is clean, validated, and aggregated.

The efficiency of an OLTP system is reduced when data is used for analysis and reporting. The data warehouse brings together selected pieces of data from multiple heterogeneous databases and information sources into a single repository suitable for querying and analysis. Information from each source that is of interest is extracted, translated, and merged with relevant information from other sources and stored in a centralized repository. When a query is posed, it is evaluated directly at the repository, without accessing the original information sources. Creating a data warehouse separate from the OLTP systems for analysis purposes ensures that the performance of the transactional system is not degraded by performing queries on them.

In a data warehouse the data not only is integrated across different functional units of the organization, but also includes external entities such as customers and suppliers. Data warehouses minimize data redundancies and eliminate inconsistencies in data. When a new information source is added to the data warehouse or when relevant information in the source system changes, it is reflected in the data warehouse. Data is integrated into the data warehouse by filtering the information, summarizing it, or merging it with information from other sources. The data warehouse also integrates data across time to provide views obtained from trend analysis of the data.

Efficiency Improvements

The importance of data warehousing rests on many aspects and results in many organizational benefits. The data warehouse provides a "single version of the truth" (Agosta, 1999; Watson et al., 2004) and better data analysis. The data warehouse can shrink the information delivery time between an event's occurrence and business decision making. Time savings and more and better information are the most tangible benefits offered by a data warehouse. Data warehouses can save time for users, facilitate the development of new applications, and provide support for customer-focused business strategies. The data warehouse empowers users as it supports end-user analytical activities (Zeng et al., 2003a).

Providing system quality is another way that data warehouses contribute to the efficiency of organizations. The data warehouse integrates data from multiple sources and flexibly supports current and future users and applications. This provides users a better understanding of the context in which they make their decisions. The data warehouse significantly increases the decision-making productivity and affects how people perform tasks by reducing time and effort.

Customer Management

A data warehouse provides the foundation to build a customer relationship management (CRM) strategy. Concepts of mass production and mass marketing have given way to new ideas in which customer relationships are the central business issue. Tyagi (2003) contends that a successful business will be one that understands its customers' behavior at the most granular level. CRM involves capturing customer data, consolidating, integrating, analyzing data, and using the results to respond to the current and potential needs of the customer (Praskey, 2001). The payoffs from a CRM include not only acquisition of new customers (Dyche, 2002), but also increased customer retention (Berson et al., 2000) as well as revenue from existing customers (Swift, 2000).

Data warehousing and data mining have made CRM a new area where firms can gain a competitive advantage. Businesses today are increasing customer value through analysis of the customer life cycle. Given the high cost of acquiring customers, marketing focus has shifted away from expanding the breadth of customer base to the depth of each customer's needs. Through data mining and the extraction of hidden predictive information from data warehouses, organizations can identify valuable customers, predict future behaviors, and make proactive, knowledge-driven decisions. Data warehousing and data mining help turn customer data into customer profiling information and provide customer value, customer targeting information, customer rating, and behavior tracking (Figure 2.6).

Data warehouses assist in customer management by allowing integrated knowledge of the customers to be stored and analyzed. Customer behaviors, buying

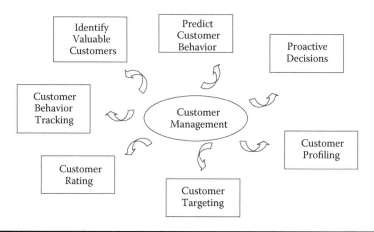

Figure 2.6 Customer management.

habits, attitudes, expressed needs, and value positions are filtered and stored in data warehouses. Data warehouses help leverage customer knowledge in product design, distribution channel decisions, and every interaction with the customers. It allows development of strategies to retain high-value customers and focus on customer needs and preferences.

Data warehouses also help in building strategies to extend relationships with present customers and target new customers. Collecting customer demographics and behavior data makes precision targeting possible. Targeting helps in devising an effective promotion plan to meet tough competition or identify prospective customers when new products appear.

Conclusion

This chapter highlights the importance of a data warehouse to an organization. It shows that a data warehouse can have a strategic as well as long-term value for an organization. However, building a data warehouse is a complex process. Designing a data warehouse requires techniques completely different from those adopted for operational information systems (Chaudhuri and Dayal, 1997; Golfarelli and Rizzi, 1998). The complexity of its development and implementation lies in the unique features of a data warehouse. The features of the data warehouse are described in the next chapter along with the differences between a data warehouse and traditional operational systems. The data warehouse components and development process are described in the following chapters.

Chapter 3

Difference Between Data Warehouses and Traditional Operational Systems

Features of a Data Warehouse

Over the years various researchers have refined the definition or broadened the scope of the data warehouse definition, but the essential characteristics remain the same. According to Ester et al. (1998), a data warehouse is a collection of data from multiple sources integrated into a common repository and extended by summary information for the purpose of analysis. Similarly, Rundensteiner et al. (2000) find that data warehouses have emerged as one key technology for the integration of distributed information sources. Gardner (1998) defines data warehousing as "a process, not a product, for assembling and managing data for the purpose of gaining a single, detailed view of part or all of a business." The main features of a data warehouse are summarized here.

Subject Oriented

In a data warehouse, data is organized according to subject instead of applications (Chaudhuri and Dayal, 1997; Gardner, 1998; Tryfona et al., 1999). A subject area identifies and groups processes that relate to a logical area of the business. In a data warehouse, the information from across functional departments or business units is organized in a manner that is subject oriented, with an enterprise view. This subject-oriented detailed transactional data allows corporate users to drill down into the depth of their business operations for data mining and business intelligence activities.

The operational environment focuses on the day-to-day operations of the business. In the data warehouse, the data is oriented differently. It is concerned with the things that drive the transactions; for example, customer, product, employee, accounts, flight, purchase, or billing. Each of these subject areas is physically implemented as several related tables in the data warehouse. A particular subject may be involved in different types of transactions. For example, a customer appearing in the accounts payable system may also be a parts supplier in the supplier system and therefore appears in both systems.

Integrated

The warehouse contains integrated data about a particular subject instead of the ongoing operations of the organization (Debevoise, 1999; Inmon, 1996a; Rahm and Do, 2000). Data is integrated as the data moves from operational systems into the data warehouse. In a data warehouse the data not only is integrated across different functional units of the organization but also includes external entities such as customers and suppliers. For example, feeds from the stock market may be integrated with financial data from operational systems in a data warehouse for a comprehensive financial analysis.

Because data warehouses are targeted for decision support, they contain consolidated data rather than detailed, individual transactional records. Data in the warehouses is integrated from several operational databases, over potentially long periods of time into one repository. Data is integrated to support a corporate view of the data. Integration is not the mere gathering of data into a single large database. Integration of data requires several processes. The two most important processes are data transformation and data cleansing.

Nonvolatile

Data in the data warehouse is nonvolatile. Once the data enters the data warehouse, it remains unchanged. In an operational system, data can be changed by deleting or modifying it. The data in the data warehouse is not updated. Any change to the information is done by adding a new record to reflect the changed status of the data. The existing records are not modified. For example, say a person's contact details are stored in the customer database as Record No. 1. In an operational system, if the person's telephone number changes, this change is made to the Record No. 1 in the customer database by modifying the entry. However, in a data warehouse no change will be made to Record No. 1. Instead, a new record (Record No. 2) will be created and inserted into the data warehouse to reflect the changed telephone number. The warehouse data is nonvolatile in that the data that enter the database are rarely, if ever, changed.

Time Variant

A major strength of the data warehouse lies in the time variance of its data (Pedersen and Jensen, 1998; Han et al., 1998). The value of the operational data archived in the data warehouse is a function of time and changes on the basis of time. A data warehouse gives an accurate picture of operational data for a given time, and changes in the data in the warehouse are based on the time-based changes in operational data. The data from the operational systems is extracted at a specific moment in time, creating a snapshot of the data. The data warehouse consists of snapshots of the operational data taken at intervals of time. Data can be viewed in the data warehouse across the field of time in different levels of detail. This time variant characteristic of the data warehouse allows complex analysis along the time dimension, allowing patterns and trends to be viewed over time.

Historical

Unlike operational systems that require real-time views of data, data warehouse applications generally deal with long-term, historical data (Chaudhuri and Dayal, 1997; Ballou and Tayi, 1999; Gardner, 1998). Data warehouses generally contain a greater volume of more detailed information over a longer period of time (Zeng et al., 2003a). They contain both atomic (Gardner, 1998; Chaudhuri and Dayal, 1997) and summarized (Widom, 1995) data. Atomic data (such as source data or raw data) is data that has not been processed for meaningful use. Summary data represents data that has already been aggregated (for example, a summary table containing total sales by product by year). The storage and manipulation of summary data reduces the amount of processing required by a query. The historical data in the data warehouse enables detecting trends and making predictions which cannot be done in an operational database because it stores only current data. Historical information allows an understanding of not only the seasonality of business but also long-term patterns of behavior as well.

Difference Between Data Warehouses and Traditional Operational Systems

Data warehousing is a collection of *decision support* technologies aimed at enabling the *knowledge worker* (executive, manager, analyst) to make better and faster decisions. Decision support places some rather different requirements on the data

Table 3.1 Comparison of Data Warehouses and Operational Systems

Data Warehouse Systems	Operational Systems
Used by management	Used by front-line workers
Strategic value	Tactical value
Supports strategic direction	Supports day-to-day operation
Used for online analysis	Used for transaction processing
Subject oriented	Application oriented
Stores historical data	Stores current data only
Unpredictable query pattern	Predictable query pattern

Source: Adapted from Sperley, E. (1999), Enterprise Data Warehouse, Upper Saddle River, NJ: Prentice Hall.

warehouse and it differs considerably when compared to traditional online transaction processing applications in operational systems.

Table 3.1 presents a comparison between the characteristics of a data warehouse and those of an operational system.

Used by Management

The data warehouse is a decision support tool used by management (Chau et al., 2003; Chaudhuri and Dayal, 1997; Wixom and Watson, 2001) mainly for analytic processing, unlike operational systems (Sumner, 2000) used for transaction processing (Jiang et al., 2000). Operational and analytic data are separate, with different requirements and different user communities. The data warehouse is maintained separately from the organization's operational databases. Operational systems are designed and optimized to handle the transactions of running the business (Klenz, 2001). Functionally oriented operational data is used to satisfy the immediate functional processing requirements of the business user.

Operational systems do not analyze data to allow higher level decisions to be made, as in a data warehouse. Data warehouses supply information to managers in the form of periodic reports (L. Chen et al., 2000) to support decision making. Decision support usually requires consolidating data from many heterogeneous sources; these might include external sources into the data warehouse.

Strategic Value

The data warehouse has strategic value for information management and decision support in organizations (Cooper et al., 2000; B. Shin, 2002; Watson et al., 2002), such as gaining competitive advantage or customer relations advantage. Data exploitation through the data warehouse can expose hidden trends in enterprise data that can be used to realize new business opportunities. Data warehousing, when properly implemented, can reduce business complexity, help organizations discover ways to leverage information for new sources of competitive advantage, and provide a high level of information readiness to respond quickly and decisively under conditions of uncertainty. Operational systems, on the other hand, have tactical value in the organization (Schmidt, 2000; Shapiro, 2001; Talluri, 2000), for example, cost efficiency, accuracy, ease of processing order entry, or speed up or reduce paperwork.

Data warehouses gain informational, as opposed to operational, access to corporate data. The data warehouse is built to leverage the investments already made in legacy systems, allowing business users to effectively move from traditional corporate data access to informational access.

Data warehouses offer organizations the potential for greater exploitation of informational assets. Informational access through data warehouse applications brings a wide variety of tangible and intangible values to the organization. It leads to a rich blend of qualitative and quantitative benefits, including a reduction in labor time, more effective and efficient decisions, and higher morale among users.

Strategic Direction

Data warehousing applications are used to redesign business processes and support strategic business objectives. The fundamental business driver behind data warehousing is the desire to improve decision making and organizational performance. These benefits involve significant changes to the way the organization operates.

Whereas operational systems handle day-to-day workings of the business, a data warehouse supports the strategic direction (Cooper et al., 2000; Watson and Haley, 1998) of the organization. A data warehouse can be a critical enabler for a major shift in an organization's strategy. This could involve reengineering a particular business process or include a coordinated collection of redesigned processes, all working together to support a new organizational strategy. Data warehouses used to reengineer processes or support the strategic objectives of the organization incur benefits like reduced cost of operations and improved information access.

Online Analytical Processing

The operational database which continuously produces operational data is based on online transaction processing applications, whereas a data warehouse supports online analytical processing applications (H. Lee et al., 2001). Online transaction processing systems address the operational data needs of an organization. However, they are not as well suited for supporting decision-support queries as data warehouses are.

Decision-support queries involving analytics including aggregation, drilldown, and slicing-and-dicing of data are better supported by online analytical processing (OLAP) systems. Data warehouses support OLAP applications by storing and maintaining data in a multidimensional format. Data in the data warehouse is extracted and loaded from multiple transactional data sources using extract, transform, and load (ETL) processes.

OLAP enables the data warehouse end user to gain insight into data through fast, consistent, and interactive access to a wide variety of possible views of information that has been consolidated and transformed from raw data (Vassiliadis and Sellis, 1999).

The concept of normalizing data in a transaction system is not applicable to data warehouses (Gray and Watson, 1998). A data warehouse continuously produces analytical information for business users. Because search and analysis efficiency is more important in a data warehouse, data warehouses are not fully normalized and contain derived or calculated data that would not be included in a transaction-based database (Moody and Kortink, 2000). Also unlike traditional operational systems, the data in a warehouse is typically modeled multidimensionally to facilitate complex analyses and visualization.

Subject Oriented

In a data warehouse, data is organized by subject areas (customer, finance) across the enterprise rather than on an application-by-application basis. In an operational system, data is organized around functional organizations within a business, to satisfy the immediate functional processing requirements of the business (Gardner, 1998). A data warehouse, on the other hand, contains data oriented to decision making. A data warehouse stores data that is subject oriented with an enterprise view, integrating information from across functional units (Gardner, 1998). A data warehouse collects information from multiple systems and stores it in a fashion that allows end users to have faster, easier, and more flexible access to key information. Operational data requirements relate to the immediate needs of the application and are based on current business rules. In a data warehouse the data span time and allow for more complex relations.

Historical Data

A data warehouse stores mass volumes of historical data in its database system for fast analysis and reporting (Anton, 2000; Berndt et al., 2003), unlike operational systems which contain current transactional data (Chau et al., 2003; Moody and Kortink, 2000). Understanding trends or making predictions requires historical data in the data warehouse.

In most organizations, the legacy systems house huge amounts of operational data as well as archived data. This historical data is made available through the data warehouse for analysis. The data warehouse makes this historical data valuable by making it available for querying and turning the detailed atomic data into useful information. The data warehouse allows detailed ad hoc analysis of the historical data by capturing the detailed data and providing access to it. Data warehouses store huge amounts of data, far beyond what is held in operational systems, and the data is kept and used for long periods of time.

Operational systems store only current data because they typically automate clerical data processing tasks such as order entry and financial transactions, which are the day-to-day operations of an organization. These tasks are structured and repetitive. The transactions require detailed, up-to-date data, and read or update records accessed typically on their primary keys.

Data warehouses are built to provide decision support and contain historical, summarized, and consolidated data, unlike the detailed individual records contained in operational systems. Because data warehouses contain consolidated data from several operational databases over potentially long periods of time, they tend to be much larger than operational databases.

Unpredictable Query Pattern

The data warehouse can support unpredictable queries (Moody and Kortink, 2000), unlike an operational system that employs predictable query patterns. The OLAP system in a data warehouse provides an information structure that allows an analyst to have flexible access to data, to slice-and-dice data in numerous ways, and to dynamically explore the relationship between summary and detailed data (Hristovski et al., 2000).

In an operational system, the consistency and recoverability of the database are critical, and maximizing transaction throughput is the key performance metric. Executing complex analytical queries against the operational databases would result in unacceptable performance of the operational system. On the other hand, data warehouse environments are query intensive, with ad hoc, complex queries that access millions of records. Query throughput and response times are more important than transaction throughput in a data warehouse because several joins and aggregates are executed while performing the queries.

A data warehouse allows the execution of cross-functional queries that are not always possible in an operational system. Data warehouses allow the integration of data from different aspects of the business to analyze trends and correlations. This contributes infinite value to decision makers. Such cross-functional querying is not possible in operational systems modeled along functional lines.

Other Important Differences

As discussed throughout this chapter, a data warehouse differs in many ways from a traditional operational system. The data warehouse supports management decision

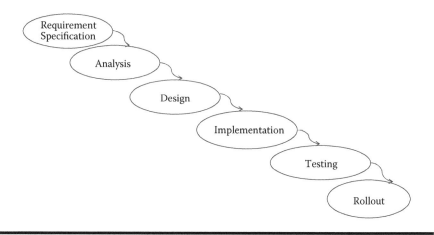

Figure 3.1 Waterfall software development cycle.

making and plays a role in supporting the strategic direction of the organization. It also integrates enterprise-wide data and provides sophisticated analyses of the data (J. Srivastava and Chen, 1999), facilitating business understanding (Inmon, 1996a; Sullivan, 2001). Therefore, it appears that for an effective implementation of the data warehouse in an organization, aligning the data warehouse to business goals and strategies would be an appropriate and necessary measure.

The day-to-day management of the data warehouse is also different from the management of an operational system (Chaudhuri and Dayal, 1997), because the volumes can be much larger (Chaudhuri and Dayal, 1997; Gardner, 1998) and require more active management (Hammer et al., 1995), such as creating or deleting summaries or rolling data on or off the archive. In essence, a data warehouse is a database that is continually changing to satisfy new business requirements (Gardner, 1998).

In practice, data warehouses must be designed to change constantly to adapt to changes in the business arena (Armstrong, 1997). In order to provide this flexible solution, Anahory and Murray (1997) have found that the process that delivers a data warehouse has to be fundamentally different from a traditional waterfall method (Figure 3.1). The waterfall method is a sequential software development method in which development flows downward (like a waterfall) through the phases of requirement specification, analysis, design, implementation, testing, roll out, and maintenance. The underlying issue with data warehousing projects is that it is very difficult to complete the tasks and deliverables in the strict, ordered fashion demanded by a waterfall method. This is because the requirements are rarely fully understood and are expected to change over time (Mohania and Dong, 1996; Strong et al., 1997).

For a data warehouse implementation strategy, Inmon (1996a) advises against the use of the classical waterfall approach. He advocates the reverse of a system

Figure 3.2   Spiral software development.

development life cycle approach such as the waterfall. He suggests that data warehouse development be driven by data instead of starting from requirements gathering.

For enterprise-wide data warehouse development, it is impractical to determine all the business requirements a priori, so the waterfall approach is not viable. To elicit the business requirements, an iterative (spiral) approach such as prototyping is usually adopted (Figure 3.2). For individual data marts, on the other hand, a phased development approach can be used such as business dimensional life cycle because they focus on business processes, which are much smaller in scope and complexity than the requirements for an enterprise-wide warehouse.

The development process of the data warehouse and its complexity is discussed in the next chapter.

Chapter 4

Data Warehouse Development Process

The core business processes of many organizations are becoming more dynamic and complex because of globalization and evolving technology (Landry et al., 2004). Agosta (1999) asserts that data warehousing is a system architecture, not a software product or application. Similarly, Manning (1999) believes that the data warehouse was intended to provide an architectural model for the flow of data from operational systems to decision support environments. Building a data warehouse requires the integration of many tasks and components and coordination of the efforts of many people (Kimball, 1998). A number of researchers (Murtaza, 1998; Meyer and Cannon, 1998) have identified various data warehousing components and dimensions. However, these dimensions are often overlapping. The following section organizes the essential components for defining and understanding data warehouses and supports a methodical approach to presenting the data warehousing process.

A data warehouse can be categorized into six major components as

1. Data sourcing
2. Data conversion and extraction
3. Data warehouse database management system
4. Data warehouse administration
5. Business intelligence tools
6. Metadata

These data warehouse components and their strengths are discussed in this chapter. Figure 4.1 illustrates the overall architecture of a data warehouse by identifying the major components and how data flows through the system.

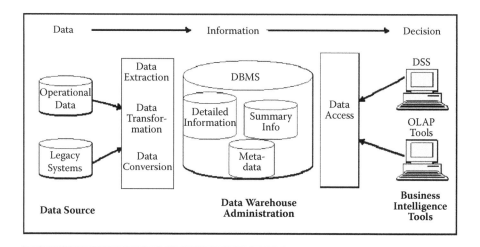

Figure 4.1 Architecture of a data warehouse. (DBMS: database management system; DSS: decision support system; OLAP: online analytical processing)

Data Sourcing

Building a data warehouse is a complex and lengthy process. First, the information needs of the organization have to be identified. This in turn helps to determine the data requirements that fulfill these information needs. These requirements are used to develop a data model that provides business reasons for building a data warehouse (Little and Gibson, 2003). Sources of data are then identified in the transactional legacy systems. Information from each source is extracted, translated, and filtered as appropriate and merged with relevant information from other sources before being stored in the data warehouse.

Information from the native format of the source is translated into the format and data model used by the warehousing system. When a new information source is attached to the warehousing system, or when relevant information at a source changes, the new information is propagated to the data warehouse. When a query is posed, the query is evaluated directly at the data warehouse repository, without accessing the original information sources.

Data Modeling

A model is the reflection of the real world in an abstract manner. Data models provide the user with the capability to visualize the abstract data. It allows the user to understand and navigate the data structure of the data warehouse and fully exploit

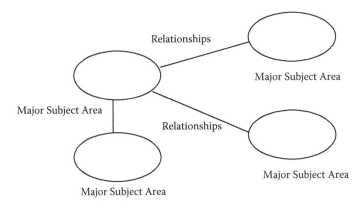

Relationships

Major Subject Area

Major Subject Area

Relationships

Major Subject Area

Major Subject Area

Figure 4.2   High-level data model.

the data. The data model helps to organize the structure and contents of the data in the data warehouse.

The corporate data model is a conceptual data model and provides an abstract representation of the information requirements of all or part of an organization. It is independent of functional boundaries within an organization and of implementation technology. A conceptual data model is a high-level model of information requirements within an organization (Figure 4.2).

Data is often duplicated throughout the organization, resulting in potentially inconsistent data that may be stored in different formats and is difficult to consolidate. Conceptual data models integrate information from a number of sources when designing cross-functional information systems and help in explaining and visualizing data in order to facilitate stakeholder understanding. The enterprise-wide data model provides a basis for communication about information in an organization and a framework for developing an inventory of data in legacy systems.

The data modeling process allows the elicitation and capture of knowledge about design decisions, assumptions, and the details of how particular stakeholders intend to use the data represented in the conceptual data model. Modeling data warehouses is a complex task focusing on internal structures and implementation issues. Capturing and explaining information requirements using scenario-based analysis and using graphical icons for visualization enhances the capture and integration of stakeholder viewpoints into the conceptual data models.

The enterprise data model and the corporate data model are good places to start the process of building a data warehouse model. The enterprise data model has to undergo some transformations to build a data warehouse data model. The enterprise data model usually identifies the major subject areas of the enterprise,

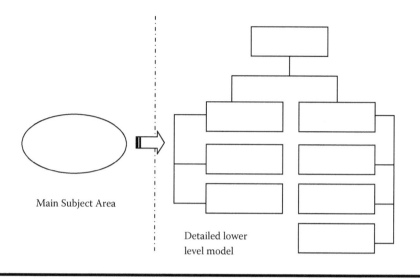

Main Subject Area

Detailed lower
level model

Figure 4.3   Subject areas have lower level models.

the relationships between the subjects, the definitions of the subject areas, and the conceptual or logical data models for each subject area. The scope of the data warehouse data model is determined by the scope of integration. The scope of integration defines the boundaries of what entities to include or exclude in the data warehouse model and needs to be defined before the data modeling process starts.

In a corporate model, groups of entity and relationship types may be clustered into high-level subject areas. The subject areas are linked by shared entity types and may overlap. Subject areas allow for structuring of corporate data models to improve their accessibility. A number of high-level subject areas may be used to represent a corporate data model, each of which may be further expanded into a more detailed model (Figure 4.3).

Conceptual data models are designed using entity relationship (ER) notation. ER modeling produces a high-level conceptual model of the data warehouse. It uses two basic concepts — entities and relationships — to model the data. The properties of the entities and the relationships are described by their attributes. A domain encompasses all possible values for an attribute. An entity represents a class of objects (person, place, thing, or event) that can be observed and classified according to its properties or characteristics. An entity usually has a clear boundary definition. In a high-level conceptual model, an entity can be more generic versus a more specific entity in a logical model. A relationship is represented with a line connecting two entities. It depicts the association between the entities in a model. In lay terms, if an entity is a noun, the relationship is the verb. The relationship between entities is defined in terms of cardinality. Cardinality could be one-to-one, many-to-one, or many-to-many.

The *multidimentional model* enables data to be easily and quickly viewed from many perspectives or dimensions. In a highly normalized data structure, performance would be impacted because several joins would be required between the tables holding the different dimension data. Multidimensional analysis allows the user to explore several interdependent factors in a business problem and view the data in complex relationships. The complex relationships can be analyzed iteratively by drilling down to lower levels of detail or by rolling up to higher levels of summarization and aggregation. For example, total home loans for a bank can be drilled down to view the loans by region, branch, and then customer. Or the analysis could start with a customer home loan and then be rolled up from branch-level total home loans to regional total home loans and further to total nationwide home loans.

Dimensional modeling uses basic concepts comprising measures, facts, and dimensions. Facts represent atomic information elements in a multidimensional model. A fact consists of quantifying values stored in measures (Husemann et al., 2000). In other words, a fact represents a business measure. The primary table in a multidimensional model is the fact table. It stores the numerical performance measurements of the organization. The measure is qualified through dimension levels. A list of dimensions defines the scope of the measurement or the grain of the fact table. Each dimension level consists of a set of elements or instances. Dimensional tables contain the descriptions of the business. They consist of attributes or columns that describe a row in the dimension table. Dimensions offer a way to organize and select data for retrieval, analysis, and exploration. Common dimension levels for aggregating data in the data warehouse are temporal (year, month, week), geographical (country, state, region), or the hierarchical levels within an organization (department, program). A fact table has composite keys with two or more foreign keys that match the primary keys of the corresponding dimensional table related to it. This maintains the referential constraints within the model.

Fact Table	Dimension Table
Order Id	Customer No
Customer No	Customer Name
Product No	Customer Birth Date
Quantity Ordered	Customer Address Line 1
Unit Price	Customer Address Line 2
Total Order Value	Customer Contact Number
	Customer Status
	Customer From Date
	Customer Preference

Several schemas are used to provide some level of modeling abstraction that is understandable to the user. The star and the snowflake schemas are used most widely. The star data structure is so named because its representation depicts a star with a central fact table joined to several outlying structures of data or dimensions that describe one important aspect of the fact table. The star structure is popularly used for the *multidimensional* view of data in the warehouse. In a multidimensional data model, a set of *numeric measures* that are the objects of analysis depend on a set of dimensions which provide the context for the measure. In other words, the multidimensional data views a measure as a value in the multidimensional space of dimensions. Each dimension is described by a set of attributes. The attributes of a dimension may be related via a hierarchy of relationships. Some key features of the multidimensional conceptual model are *aggregation* of measures by one or more dimensions and *comparing* two measures aggregated by the same dimensions. In multidimensional models, the time dimension is of particular significance because it allows for trend analysis.

Star schemas are further refined into snowflake schemas. The snowflake model is anchored by a central fact table with dimension tables surrounding it. It allows for attribute hierarchies by having subdimensional tables under the dimension tables of the star schema. When necessary, it provides for logical separation of data. In the snowflake data model, more than one fact table is combined to create a composite structure similar to a snowflake. In this structure, different fact tables are connected by sharing one or more common dimensions or conformed dimensions. The drawback of the snowflake schema is that the subdimensions generally slow down the query processes by adding complexity to the schema.

Transforming Enterprise Data Model to Data Warehouse Model

To transform the corporate or enterprise data model into a data warehouse model, several steps must be taken (Inmon, 1993):

1. Removal of purely operational data
2. Addition of an element of time to the key structure of the data warehouse
3. Addition of appropriate derived data
4. Transformation of data relationships into data artifacts
5. Accommodation for different levels of granularity found in the data warehouse
6. Merging of like data from different tables
7. Creation of arrays of data
8. Separation of data attributes according to their stability characteristics

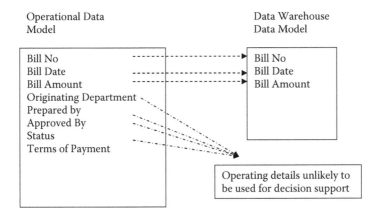

Figure 4.4   Removing operational data.

Remove Operational Data

The operational system captures all the data that is used for day-to-day running of the business. However, some of this data may not always be used for decision support purposes. To convert an operational corporate data model into a data warehouse data model, data that is used purely in an operational environment is removed (Figure 4.4). Situations can arise where all the data may be used for analysis purposes. Carrying over all operational data to the warehouse environment involves huge volumes of data and is very costly. A decision needs to be made to balance the reasonableness of the use of the operational data for decision support versus its use in exceptional or farfetched cases.

Add an Element of Time to the Key Structure

The data warehouse contains data that is time-variant. It is captured at a snapshot of time. This element of time is added to the key structure of the data warehouse. The element of time can be added in various ways. One popular method is to add it in the form of a snapshot date and time (Figure 4.5). Another method is to add a *from datetime* through a *to datetime* in the key structure. The advantage of this method is that it allows for the capture of continuous data rather than snapshots of data in time.

Add Appropriate Derived Data

The data warehouse is built to support decision making. Derived data is added to the data warehouse model where the derived data is calculated once rather than

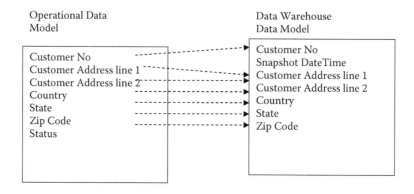

Figure 4.5   Adding an element of time.

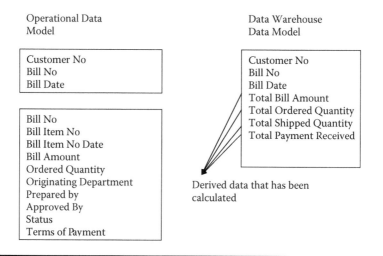

Figure 4.6   Adding derived data.

repeatedly. Addition of derived data reduces the processing time required for accessing data in the data warehouse. Another advantage is that once the derived data in the form of calculated and summarized data is calculated and stored in the data warehouse, it does not have to be calculated every time it is needed. This reduces the chances of data being derived differently by different users and increases the integrity and credibility of the derived data in the data warehouse. (See Figure 4.6.)

Transform Data Relationships into Data Artifacts

A data warehouse contains data over a period of time. The data relationships captured in the data warehouse are different than those that underlie an operational system. In an operational system there is usually only one business rule underlying

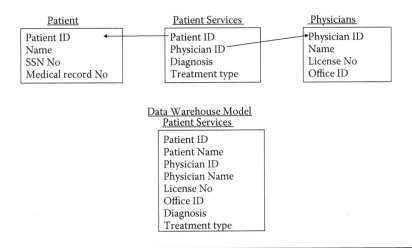

Figure 4.7 Artifacts of data relationships.

a data relationship. In a data warehouse, there are usually many relationship values between tables of data, because it represents data over a long period of time. These relationships between tables in the data warehouse are captured by creating relationship artifacts.

Relationship artifacts show the existence of the relationship at the snapshot of time that the data was captured into the data warehouse. The artifact may include foreign keys and relevant data from the associated table. For example, Figure 4.7 shows the relationship between a patient and a physician in a healthcare group practice. Each patient has a primary care physician. Integrity constraints dictate that if a physician leaves the group practice and is deleted from the provider list, a provider services record may not exist that has no physician as the primary source. The information about who provided the service to the patient will be lost if the provider record is deleted.

In the data warehouse this information is captured through transforming the data relationships into data artifacts. Referential integrity appears as artifacts of relationships in the data warehouse environment. For example, the patient services table will capture more detailed information when the snapshot of data is taken. One of the pieces of information that would be captured is the primary physician name at the moment of the snapshot. Another artifact that may be captured is the physician license number or disease diagnosis from the operational system.

Accommodate Different Levels of Granularity

Granularity refers to the level of detail or summarization of the units of data in the data warehouse. As the level of detail increases, the level of granularity in the data warehouse decreases (Figure 4.8). For instance, every customer activity record

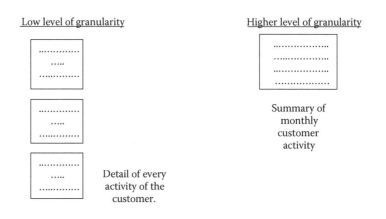

Figure 4.8   Different levels of granularity.

would be at a low level of granularity. A summary of customer activity for a monthly or biweekly period would be at a higher level of granularity.

The data warehouse provides the foundation for enterprise-wide decision support. When the data warehouse is properly designed and constructed, it provides a flexible and reusable foundation to support various types of decision support processing. The granularity of data found in the data warehouse provides reusability of the data because the same data can be used by different users in different ways. For example, in a healthcare environment, the finance department may use data for the total patients admitted per month for monthly revenue purposes, physicians may use the data for total patients admitted to study the total occurrences of a disease by type of disease incidence, and catering may use the same data for total patients admitted to evaluate and project their food purchases. Looking at the same data in different ways is the advantage offered by the data warehouse.

Low levels of granularity in the data warehouse allow for flexibility to analyze data in great detail. When the data is in a low enough level of granularity, it can be reshaped to meet many different needs and demands across the organization. Another advantage of data with low levels of granularity is the ability to be reconciled. If one needs to explain discrepancies of results in two different departments, the data warehouse data, on which everyone relies, can be used to analyze and explain the discrepancy. Yet another advantage offered by low levels of granularity of data is the ability to meet new requirements. The data warehouse data allows response to change, by making data available for analysis. For example, to see the effect due to a change in external policy or tax law changes, the information in the data warehouse can be used to study the effects brought about by the new changes.

The data warehouse stores different levels of granularity of data, determined by user requirements. If the volume of data is large and there is an issue of cost, a high level of granularity is more efficient for storing data. Higher granularity of data allows compacting of data and requires lower disk space, lower processing power

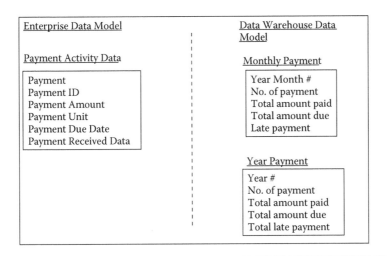

Figure 4.9   Granularity of data.

to access the data, and fewer indexes to be built. The tradeoff is a lower ability to answer any type of query. For example, in a department store data warehouse with a low level of granularity of data, a customer's purchase on a certain day for a certain item can be determined, because every sale was recorded. On the other hand, with a high level of granularity, where only a monthly or weekly summary of purchases is recorded in the data warehouse, there is no definitive way to determine a particular purchase by the customer.

To overcome this problem, most data warehouses store dual levels of granularity: lightly summarized data and true archival data that is at the detailed level. Most of the query processing occurs at the lightly summarized data level. The dual levels of granularity in the data warehouse offer efficiencies of cost and access and also allow the ability to answer any question.

In the data model, the hierarchies of a dimension show the different granularities at which the connected fact set can be summarized (Figure 4.9). For example, the sum of the repayments per customer on their installments can be summarized along the dimension of time on a quarterly basis or on a monthly or weekly basis. The levels of granularity of the data stored in the data warehouse depend on several factors, such as the period of time used for summarization, the elements of data required for summarization, and the tradeoff between lower levels of granularity for detailed analyses versus cost of storing details.

Merge Like Data from Different Tables

The corporate or operational tables are frequently merged into a common data warehouse table to enhance the querying and analytical ability of the data warehouse (Figure 4.10). As tables are loaded into the data warehouse they are often

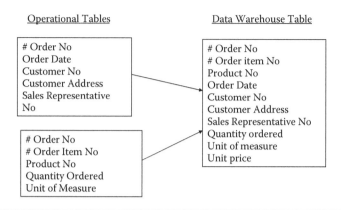

Figure 4.10   Merging tables.

merged together. Merging operational tables simplifies the data structure in the data warehouse and helps avoid commonly required joins. Merged tables are beneficial when the operational tables share a common key and the data from the merged tables is used together frequently. Otherwise, merging of tables is not necessary. For example, in Figure 4.10 the common element in the work order tables from the operational system is "# Order No." In this instance the data from the two tables is used together. There would be no order item numbers without the order numbers. As they enter the data warehouse environment, the two operational order tables are merged together, for easier query and reporting based on "# Order No.".

Create Arrays of Data

Data warehouses contain large volumes of data. To answer queries efficiently requires highly efficient access methods and query processing techniques. The creation of arrays of data in the data warehouse model enhances the performance of the data warehouse. One of the aspects of a data warehouse is the inclusion of an element of time in the key structure. Because the data warehouse contains large volumes of data, creating arrays of data along the time dimension is particularly useful.

Data warehouses store selected summary tables containing pre-aggregated data along the dimension of time by creating arrays of data in the data warehouse data model. Unlike the enterprise data model, the data warehouse data model is not completely normalized. The data warehouse data model contains repeating groups of data in the form of arrays of data. For example, sales data is captured in the enterprise data model on a monthly basis. When this data enters the data warehouse, it is organized into an array where each monthly record is an occurrence of the array for the year. The individual record for each month is stored in one structure for each year. This not only reduces storage space in the data warehouse

Enterprise Data Model		Data Warehouse Data Model
		Sales representative ID
January	Sales representative ID	Year
	Total sale amount for	Total sale amount
	month	January sales – amount
	Year and Month	February sales – amount
		December sales - amount
February	Sales representative ID	
	Total sales amount for	
	month	
	Year and month	
December	Sales representative ID	
	Sales amount for month	
	Year and month	

Figure 4.11   Arrays of data.

but also allows for speedy access during querying. Creating an array of data in the data warehouse, in this example, reduces the index entries to one twelfth — one for each year in the data warehouse model versus an index for every month of the year in the enterprise data model structure — thereby speeding up query performance times. (See Figure 4.11.)

Separate Data Attributes by Stability Characteristics

Stability analysis is the final design activity in transforming a corporate data model to the data warehouse data model. Stability analysis involves grouping attributes of data together based on their propensity for change (Inmon, 2005). In the operational environment, a table may contain some data attributes that rarely change, some data attributes that change sometimes, and other data attributes that change much more frequently (Figure 4.12). In the data warehouse model, the data is grouped separately depending on its stability characteristics. Organizing data in the data warehouse according to its stability characteristics avoids reentry of all values for an entity into the data warehouse if a single value for an entity changes. Frequently changing values stored in a separate table avoids multiple instances of the more stable data. For example, in an airline database a single table records customer activities. It captures the customer information and the customer's flight activity. In the data warehouse, this data is grouped into three categories: (1) data that is relatively stable, like the date of birth, gender, or country of birth of the customer; (2) data that changes sometimes, like marital status, country of residence, country of citizenship, or passport details; and (3) data that changes much more frequently, like flight number, place of origin of the flight, and place of destination

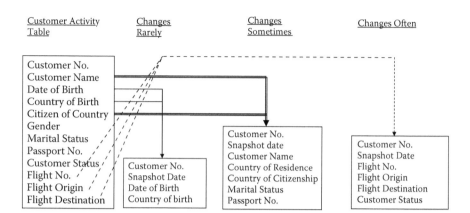

Figure 4.12 Data organized by stability characteristics.

of the flight taken by the customer. This helps reduce the data in the data warehouse because stable data, like the country of birth, or less frequently changing data, like country of citizenship, does not have to be re-recorded every time the customer takes a flight.

The logical data model provides the base for the physical database design of the data warehouse. It ensures that the data warehouse developed meets the user requirements and user expectations. The flexibility built into the design allows for handling of changes in the business rules of the enterprise over time without restructuring the database design. The physical database design implements the informational requirements of the logical data model. It takes into consideration the performance capabilities of the selected database management system. A logical data model can be implemented physically in many different ways depending on the user requirements, analytical capabilities, transaction frequencies, data volume, performance, and implementation considerations (choice of database platform).

Data Extraction and Conversion

The next step in building the warehouse is data preparation and data cleansing. It involves the extraction of source data, transformation into new forms, and loading into the data warehouse environment. As organizations realign their information infrastructure toward integrated data warehouses and decision support systems, the complex problem of accurately identifying and merging databases becomes critical (Berndt and Satterfield, 2000). According to Manning (1999), the cost of extracting, cleaning, and integrating data represents 60–80% of the total cost of a typical data warehousing project. Ensuring high-quality data is one of the most difficult challenges faced in data warehousing (English, 1999; Wixom and Watson, 2001).

Data Extraction

Data is extracted from the operational systems by extraction routines. The extracted data is then converted into an intermediate schema and placed in a staging area. The source data accumulated in the staging area is subjected to data cleansing, transformation to the intermediate schema, and data aggregation and finally loaded into fact tables in the data warehouse. Extracting data from source systems for a data warehouse can be a complicated process requiring customized code for extraction.

Data Cleaning

Because a data warehouse is used for decision making, it is important that the data in the warehouse be correct, to avoid wrong conclusions. For instance, duplicated or missing information can produce incorrect or misleading statistics. In the data warehouse environments involving multiple, heterogeneous information sources, the need for data cleansing increases significantly.

As the data comes into the data warehouse from heterogeneous sources, the extracted data undergoes data cleaning and data transformation before being loaded into the data warehouse. Data is reconciled by integrating it with different formats, values, or codes. It is validated to identify inconsistent data and is filtered according to the requirements of the data warehouse. Various data conflicts arising due to conflicting source schemas are resolved. Naming conflicts in data can arise when different source systems refer to the same data using different terminologies. The same name may denote different pieces of data, or the same entity may be referred to in different ways. Conflicts may also arise when different sources contain the same data but at different levels of abstraction.

When multiple sources are integrated, the probability of encountering copies of the same information from multiple sources represented in similar or different ways increases. Related information from multiple sources can be inconsistent due to misspellings during data entry, missing information, or invalid data. Even in the same source of information, significant inconsistent data is produced due to errors in data entry. To provide accurate and consistent data, errors and anomalies in the data have to be removed.

Data is cleaned through processes such as *data migration, data scrubbing,* and *data auditing.* To ensure high data quality, data warehouses must validate and cleanse incoming data from external sources. When multiple sources of data need to be integrated, each source may contain dirty data. Data in the sources may have different representations, be redundant, or contradict, because the data sources were developed and maintained independently to serve specific needs. Data is cleaned through consolidation of different data representations and elimination of duplicate information. Because data is migrated from different sources, a fundamental task in data cleaning is the detection of records in a database that refer to the same entity but have different representations across relations or across databases. This data is cleaned by using a text matching approach, where similar textual entries are matched together as potential duplicates. Data cleaning is also achieved by removing inconsistent field lengths, inconsistent descriptions, inconsistent value assignments, missing entries, and violation of integrity constraints.

When data is merged from different data sources into a data warehouse, the resulting heterogeneity in data models, schema designs, and the actual data is addressed by the steps of schema translation and schema integration, respectively. To provide access to accurate and consistent data, different data representations are consolidated and duplicate information is eliminated. During data migration, transformation rules are specified (for example, replacing sex by gender) to clean the data. Transcription errors, incomplete information, and lack of standard formats are also addressed during data migration and data cleaning.

Data scrubbing involves detecting and removing errors and inconsistencies from data in order to improve the quality of data. Data scrubbing involves a complex cleaning and mapping process that is the most labor intensive part of building a data warehouse. During the cleaning process, desired information is filtered out and its quality is maintained for the target system. Domain-specific knowledge is used to scrub the data. Often parsing and fuzzy matching techniques are employed to accomplish cleaning from multiple sources.

Data auditing tools make it possible to discover rules and relationships or to signal violation of stated rules by scanning data. By incorporating a set of internal controls, it enhances the system's reliability and makes it possible to prevent, detect, and eliminate data errors, irregularities, and fraud. For example, based on statistical analysis, data auditing may discover a suspicious pattern in the data of the data warehouse that shows a certain customer has returned more than the average number of goods at a particular store location.

Loading the Data Warehouse

The ETL (extract, transform, load) process transports and transfers data from the operational source systems into the data warehouse for analysis. The loading process in a data warehouse usually consists of complex specifications to manage parallel operations, and interaction between operations, on the data extracted from the source systems. Parameters such as availability of source, customization, and integration of data have to be coordinated while transferring data from source systems to the data warehouse.

After it is extracted and cleaned, the data is loaded into the data warehouse in two basic ways: a record at a time or en masse by using a utility. Additional preprocessing may be done while loading data, for checking integrity constraints, sorting, summarizing, and aggregating data to build the derived tables in the warehouse. In addition, indexes must be loaded at the same time the data is loaded. Typically, batch load utilities are used for loading the data into the data warehouse. In addition to populating the warehouse, a load utility allows the system administrator to monitor status; to cancel, suspend, and resume a load; and to restart after failure with no loss of data integrity. Because the load volumes for data warehouses are much larger than for operational systems, the load is often parallelized. Sequential loads can take a very long time, so the data is frequently divided into several job streams. Another approach used to efficiently load large volumes of data is staging the data prior to loading. Staging of data is needed when the complexity of processing is high or when there is a need to coordinate merging of data from more than one source. Scenarios are possible when data from the first source is ready for loading but has to wait in the staging area for data from another source to be available with, which it has to be merged, prior to loading into the data warehouse.

For not-so-complex scenarios, data can be loaded as a long batch transaction that builds up a new database. While the load is in progress, the current database can still support queries. The current database is replaced with the new one when the load transaction commits. Periodic checkpoints are monitored during loading to ensure that if a failure occurs during the load, the process can be restarted from the last checkpoint.

Refresh

A data warehouse is refreshed periodically to propagate updates on source data to correspondingly update the base data and derived data stored in the warehouse. The information in the sources is monitored for changes to the data that are relevant to the warehouse. These changes are then propagated to the data warehouse. The data warehouse is refreshed periodically to reflect the changes. Refresh techniques depend on the characteristics of the source and the capabilities of the database servers. Most database systems provide replication servers that support incremental techniques for propagating updates from a primary database to one or more replicas. Such replication servers incrementally refresh a warehouse when the sources change. Extracting an entire source file or database is usually expensive, but it may be the only choice for legacy data sources. Changes are notified from source systems either by providing triggers or by maintaining a log from which changes to the source can be extracted. Or, for sources that do not provide triggers, logs, or queries, periodic dumps or snapshots of the data are provided offline, and changes are detected by comparing successive snapshots.

Sometimes instead of change detection, entire copies of relevant data from the information sources are propagated to the warehouse periodically. This data is merged with existing warehouse data from other sources. In some cases the complete information from all sources is recombined and the warehouse data is recomputed.

The refresh policy depends on user needs and traffic and may be different for different sources. Ignoring change detection is acceptable in certain scenarios, when it is not important for the warehouse data to be current or when it is acceptable for the warehouse to be offline occasionally. However, when currency, efficiency, and continuous access to the data warehouse are required, then detection and propagation of changes incrementally into the warehouse is preferred. In addition to propagating changes to the base data in the warehouse, the derived data also has to be updated correspondingly. For data warehousing, the most significant classes of derived data are summary tables, single-table indices, and join indices.

The issue of frequency of refresh needs consideration when propagating changes into the data warehouse. In query able sources, if the frequency is too high, performance will degrade, whereas if the frequency is too low, changes of interest may not be detected in a timely way. The refresh cycles have to be properly chosen so that the volume of data does not overwhelm the incremental load utility. In snapshot sources, the challenge is to compare very large database dumps and detect the changes of interest in an efficient and scalable way.

Data Warehouse Database Management System

At the core of the data warehousing system lies a good data management system. The database server used for a data warehouse is responsible for the provision of robust data management, scalability, high-performance query processing, and integration with other servers (Shahzad, 1999). Warehouse servers can be categorized into two types: relational database management system (RDBMS) (Gardner, 1998; Vassiliadis, 2000) and multidimensional database (MDD) (Dinter et al., 1998).

A relational database is a type of database that stores data in the form of a set of formally described tables from which data can be accessed or reassembled in many different ways without having to reorganize the database tables. Unlike a flat file database, where data is contained in a single table, in a relational system the database is spread across multiple tables. The RDBMS supports a tabular structure for the data, with enforced relationships between the tables. The relations between the tables are also stored in the relational database, which makes it powerful for extracting data from the database. An RDBMS allows the creation, update, and administration of the relational database. The relational database offers flexibility in posing queries and viewing data in the database. Structured Query Language (SQL) is widely used to access the relational database. Oracle, DB2, and SQLServer are examples of RDBMS.

A multidimensional database (MDD) can be viewed as a cube, where information is piled on the various axes or dimensions of the cube (Buzydlowski et al., 1998; Li and Wang, 1996; Niemi et al., 2003). In a multidimensional database, data is represented in dimensions instead of relations. It presents a multidimensional view of data. Data is represented as a data cube where each individual data item is contained in a cell within the cube. In a multidimensional database, entities such as customers, locations, and sales all represent different dimensions of data. MDDs allow the definition of further hierarchies within a dimension. For example, day, week, month, quarter, and year are additional hierarchies within a temporal dimension. Typical dimensions found in data warehouses are time, organizational structures, geographic areas, and product data. The multidimensional database provides the advantages of enhanced data presentation, analysis, and navigation over relational database systems. MDDs are widely used for online analytical processing applications such as trend analysis over time or summarizations along business parameters, because they can present multiple views of data along various dimensions. It is a technique that allows multi-part questions to be posed of the database. For example, instead of a report on revenue by branch, MDD might report revenue by branch, subdivided by product lines and by region (O'Sullivan, 1996).

Multidimensional databases are frequently used to compare data over long periods of time and for computation of benchmark values from data of previous periods. Some common MDD operations are aggregation, drill down, filtering, slicing, scoping, and pivoting. In aggregation, data is summarized for the dimension the user wants to query (for example, region-wise total sales for a product). During drill down, more fine-grained data is queried by the user along a dimension. For example, for a particular zone find detailed sales, date wise made by salespersons assigned to that zone. Filtering involves screening data by evaluating it against a fixed condition, such as finding all patients in the state of Florida who have had breast cancer. A slice is a subset of selected data over a dimension of the multidimensional cube that satisfies a particular condition. Scoping is similar to subset, the difference being that all further operations of update or modification are applied only to the specified subset. Pivoting involves the rotation of the dimensional orientation of the cube. For example, in a two-dimensional array of customers versus airline, pivoting would result in a two-dimensional array of airline versus customers.

Differences between Relational and Multidimensional Designs

There are some differences between relational and multidimensional database designs for the data warehouse. RDBMS has an edge on MDD when considering its huge data storage capacity, portability issues, or security. MDD is popular for its instant response, implementation ease, and integration with metadata (Shahzad, 1999).

The multidimensional database systems directly support the way in which users analyze, visualize, and use the data. Online analytical processing (OLAP) requires analysis of very large volumes of complex and interrelated data from more than one perspective. Multidimensional database systems store data along dimensions, where each dimension represents a user perspective. This makes the MDD more efficient than RDBMS. The drawback is that restructuring an MDD is more expensive than a RDBMS.

One of the most important differences between relational and multidimensional designs lies in the level of flexibility. The relational approach to database design has an advantage over the multidimensional approach in terms of its flexibility. There is no limit to the new data that can be added to a relational model. The relational design is shaped by the abstraction of corporate data. In the relational design the data is not optimized for performance for a particular set of processing requirements. New data elements can be added to the relational design, and the scope of data is not limited. The relational design is better suited for servicing future and unknown needs.

On the other hand, the advantage of the multidimensional over relational design is its performance. It is efficiently tuned to cater to the needs of one user community. Because the database design is optimized for performance in a particular way, adding new processing requirements can affect optimization. When new requirements are added to the multidimensional design, the database design may not be able to change gracefully. Therefore, multidimensional designs are better suited for a limited scope. The optimization of data for one group of users in the multidimensional design is at the expense of performance to other groups of users who may use data in a different manner. Therefore, in terms of serviceability, the multidimensional design is good for speedy direct access of data, whereas the relational design is better for indirect access to data. In practice, most MDDs are built to complement data warehouse architectures. It gets its data from the relational database and offers it to the OLAP applications.

The relational design is much better suited to the design of an enterprise-wide data warehouse. The enterprise-wide data warehouse supports a wide base of users with different needs and perspectives of looking at the data. Because the data warehouse is not optimized for use by any one set of users, it is able to support many different uses by many different sets of users. If the performance of data for a given set of users needs to be enhanced, data can be extracted into a customized table from the relational files. This customized table can then be optimized for faster access for the single set of users. The merging of relational files into a customized table is easily done in a relational database design because data is stored in a more granular, normalized level and the relations between the relational tables are identified by foreign keys. Because relational tables can be readily reshaped another way, it is more suited for a data warehouse environment.

Data Warehouse Administration

Data warehouses present many complex administrative issues that are different from those in transactional or decision support applications (Benander et al., 2000). Data warehouse administration keeps the data warehouse environment working. With the number of subject areas and amount of historical data, a data warehouse requires significant amounts of disk storage and extensive planning (Chaudhuri and Dayal, 1997; Gardner, 1998). Data warehouse administration provides query management (Gupta et al., 1995; Widom, 1995), access control (Date, 1999; Gardner,

1998; Roussopoulos, 1998), disaster recovery (Armstrong, 1997; Sen and Jacob, 1998), tool integration (Armstrong, 1997; Freude and Konigs, 2003; Muller et al., 2000; Sen and Jacob 1998), directory management, security (Gardner, 1998; Katic et al., 1998), request control (Agrawal et al., 1997), capacity planning (Chaudhuri and Dayal, 1997), data usage auditing (Gardner, 1998; Vassiliadis et al., 2000), and user administration (Chaudhuri and Dayal, 1997; Katic et al., 1998). Effective governance is considered a key to data warehouse success (Watson et al., 2004).

The management and administration of the data warehouse, from both a technical and an organizational viewpoint, requires sophisticated and time-consuming efforts. Data warehouse administration is required for enhancing performance of the warehouse as well as for monitoring the data warehouse. Several factors influence the performance of the data warehouse. Online transaction processing issues like table access and locks encountered while accessing tables affects the performance of a data warehouse. The speed of the server affects the performance of a data warehouse. Online analytical processing issues such as monitoring queries, managing free database space, and adjusting extraction processes also affect the performance of the data warehouse. The loading, cleaning, and auditing processes in the data warehouse affect the performance of the data warehouse as well. The data warehouse database requires special methods for optimization of cache management and indexes.

Data warehouse administration is required for handling the growth issues faced by an organization's data warehouse. As the data warehouse is used, user demand for greater capabilities and for newer processing requirements increases. These demands have to be managed along with the daily and weekly processing, security, and disaster recovery planning of the data warehouse. A successful data warehouse needs to be scalable to meet user demands. As the size of the warehouse increases, computing resources need to be added incrementally without the users experiencing degradation in the performance of the data warehouse or affecting their application usage.

Access and Business Intelligence Tools

Once the data is loaded into the database, various access tools are used for end-user interaction. Gray and Watson (1998) define access tools as decision support tools that allow users to analyze information with ease. The selection of the right end-user tool is important because the ease of use and range of functions provided by the access tools determine the user's perception of the value and success of the data warehouse. These tools could be a set of query generation and reporting tools. The main goal of these tools is to remove the SQL generation of the query from the end user and make the data easily accessible (Armstrong, 1997). Or they could be more sophisticated OLAP and ROLAP (relational OLAP) tools for multidimensional analysis and data mining (Meyer and Cannon, 1998). OLAP operations roll up (increase the level of aggregation), drill down (increase details), slice-and-dice (selection and projection), and pivot (reorient) the multidimensional view of data (Chaudhuri and Dayal, 1997). ROLAP enhances query performance by selecting and materializing in summary tables appropriate subsets of aggregate views that improve overall aggregate query processing (Dehne et al., 2003; Kotidis and Roussopoulos, 1998).

Being able to consolidate and analyze data for better business decisions can often lead to competitive advantage. Business intelligence tools uncover and leverage these advantages (IBM, 1999). The most common business intelligence tool used by organizations is data mining. Mining the data warehouses provides new insights into value adding business processes, customer buying patterns, fraudulent activity, and product profitability. *Data mining* can be defined as analyzing the data in large databases to identify trends, similarities, and patterns to support managerial decision making (Zorn et al., 1999). Data mining models fall into three basic categories: classification, clustering, and associations and sequencing. Data mining allows end users direct access to and manipulation of data from within the data-warehousing environment without intervention of customized programming (Oakley, 1999). Data mining incorporates a variety of tools and processes that can work independently or together to analyze and discover relationships in collections of data (Landry et al., 2004).

Landry et al. (2004) divide data mining into primarily two types. The first, *directed* data mining, is designed to test and measure expected patterns of business behavior. The second style, *undirected* data mining, seeks patterns or relationships without any preconceived expectations or hypotheses. Data mining then lets users search large volumes of data for patterns that can be generalized in order to improve future decisions.

There is a symbiotic relationship between the activity of data mining and the data warehouse (Inmon, 1996b). A data warehouse contains integrated, summarized, and detailed as well as historical data. A data warehouse also contains metadata. Each of these elements found in the data warehouse enhances the data mining process. Integrated data saves the data miner time that would otherwise be spent on data cleansing and transforming raw data to make it usable for effective data mining. In the data warehouse, the data contained is already reconciled and the data structures are standardized. The data warehouse contains detailed as well as summarized data. Data mining processes can be applied to the granular detailed level of data to uncover patterns or at higher levels of available summarized data to save time. The historical data contained in the data warehouse allows the data miner to compute trends or to discover patterns of behavior such as studying business cycles.

Metadata

Another important component of a data warehouse is the *metadata*. Metadata is data about the data. It is data that is used to describe other data. Before a data warehouse is accessed, it is necessary to understand what data is stored in the data warehouse and where is it located.

Metadata indexes information and monitors its use (O'Sullivan, 1996). It plays an important role in the loading, organization, and utilization of data all through the data warehouse life cycle (Shi et al., 2001). Kimball (1998) define metadata as "all the information in the data warehouse environment that is not the actual data itself." Organizations need metadata for tool integration, data integration, and change management. In the context of the data warehouse, Sen (2004) describes metadata to be of two kinds: back room metadata and front room metadata. The back room metadata guides the extraction, cleaning, and loading processes. The front room metadata is more descriptive and helps query tools and report writers.

In addition to describing and locating data, metadata contains the definitions of the databases and the relationships between data elements. Metadata is data about the data and defines raw data. In an organization, different departments may be defining the same piece of information (e.g., customer number) differently. The need for metadata in the data warehouse is crucial because it contains data integrated from different parts of the organization. These differences in defining the data have to be reconciled for an application running on data warehouse data. Metadata is needed at different levels for a data warehouse. Metadata at the operational level is required for defining the data structures at the operational level; at the warehouse level, for interpreting the transformed data in the data warehouse data structures; and at the business level to map the warehouse metadata to the business concepts. Effective metadata management is required to create, manage, and map metadata in a data warehouse environment.

The data sources of a data warehouse are heterogeneous. Metadata is required about the technical characteristics, like syntax of data and location of data, as well as semantic metadata, which is needed for defining reports. Users need metadata to interpret the data in the data warehouse. When data is transformed and moved into the data warehouse, information on the source-to-target mappings of the data, transformation rules, location, and scheduling of the data is recorded in the metadata. As data is integrated into the data warehouse from multiple sources, it undergoes a scrubbing process. Transformation rules are implemented during the scrubbing process to resolve inconsistent data, incorrect data, missing data, and semantic discrepancies. Information is stored in the metadata to describe the transformations and processes performed on the data. Metadata allows the tracing of the data to its origins.

Metadata supports the user in navigating the data in the data warehouse. It plays a crucial role in the usability of the data warehouse. Metadata enables the users to search for suitable views for specific information. It also explains to the user the values of the fields. It is also the basic tool used by the administrator for the maintenance and evolution of the data warehouse. The metadata is housed in the data dictionary of the data warehouse. The data dictionary contains information relevant to the management and use of the data warehouse. The data dictionary stores the schemas of the databases as well as the corresponding mappings. Metadata consists of both the structural information of the dimensions of the data warehouse as well as the derivation logic, validation procedures, and status of the data. Metadata may contain information about the data flow in the data warehouse, for example, the direction and frequency of data feed to the data warehouse. Metadata also contains version control information, data usage statistics useful for profiling the data in the data warehouse, and security information on who is allowed to access the data.

In the last 30 years there has been a tremendous growth in the use of metadata in developing information systems, especially in the world of databases and data warehouses (Sen, 2004). But it has largely been perceived as a technology solution. Metadata management has become increasingly important in data warehousing because organizations view data as a strategic resource and data warehouses make it available for decision making. Gatziu et al. (1999) point out that only if data is linked to clear business terms, it obtains meaning for the end user of the data warehouse. Shankaranarayan and Even (2004) point out that although the benefits of metadata and challenges in implementing metadata solutions are addressed in practitioner publications, explicit discussion of metadata in academic literature is rare.

In a data warehousing environment, the end users access data directly using query tools rather than relying on reports generated by IT specialists. Because metadata empowers the data warehouse users by helping them meet their own informational needs, finding where data exists, what it represents, and how to access it (H. Lee et al., 2001), managing changes in metadata becomes important to keep the data warehouse aligned to business needs. A flexible data warehouse would remain more resilient to changes in analysis requirements over time (Moody and Kortink, 2000) and therefore better aligned to business needs and strategies.

To manage change in metadata, Sen (2004) proposes that a metadata warehouse be designed to store metadata and manage its changes for organizational decision support. Little and Gibson (2003) contend that the metadata model must be sustainable for many years to provide consistency. The metadata model should be flexible enough to provide growth of the data warehouse while consistently providing integrity for data mining and decision support systems.

A data warehouse plays an important role in integrating enterprise information. Shankaranarayan and Even (2004) contend that integrated management of metadata has not been adequately addressed by researchers. Because data warehouses aggregate vast amounts of information, integrating data warehouses with their metadata offers an opportunity to create a more adaptive information system. H. Lee et al. (2001) have proposed a metadata-oriented data warehouse architecture that consists of seven components: (1) legacy system, (2) extracting software, (3) operational data store, (4) data warehouse, (5) data mart, (6) application, and (7) metadata. They point out that metadata must be integrated with data warehousing systems because without metadata, the decision support of the data warehouse is under the control of technical users. As the data warehouse evolves, extracting data from online transaction processing systems to data warehouses becomes more complex. If metadata is integrated with data warehouse, the extraction can be automatic.

Chapter 5

Data Warehouse Architectures

The data warehouse has emerged as a key technology for integration of distributed information sources within an organization (Rundensteiner et al., 2000). Because data warehouses support the strategic direction of an organization, an architectural choice for building the data warehouse driven by business strategy would align the data warehouse more closely to the business strategy and goals. A range of architectural approaches is available for building a data warehouse. According to Murtaza (1998), the scope of the business vision can dictate the architectural approach. A short-term vision would require a lower budget, quick return on investment, and implementation with a small resource requirement, as offered by data marts. More strategic objectives of long-term gain and full organizational control would necessitate the enterprise data warehouse architecture (Murtaza, 1998). Long-term success needs a forward-thinking approach that aligns data warehousing technology with an organization's strategic objectives (Weir et al., 2003). The most popular architecture choices outside the enterprise warehouse model are the operational data store, the decision support system (DSS) data warehouse, and the data mart. These alternatives and the enterprise model itself are discussed in this chapter.

In an *enterprise data warehouse*, all business data from different operational and functional departments of an enterprise are integrated and stored in a single database with a single enterprise model (Zhou et al., 2000). It consists of a huge subject area and vast amounts of different operational sources.

An *operational data store* is a rudimentary data store that provides a collected, integrative (Inmon, 1993) view of volatile transactional data from multiple operational systems (Samos et al., 1998). An operational data store contains subject-oriented, integrated data that is current or near current to support day-to-day operational decisions. An operational data store is a repository of analysis data that has low granularity and a short retention period to keep the size of the data store manageable. The data in the operational data store is volatile and is updated. An operational data store allows an organization to do online analytical processing (OLAP) analysis without venturing into a full-blown data warehouse. Cross-system referencing requires data from more than one system. Sharing of data from more than one system leads to consistency problems when integrating data from more than one system. The risk with an operational data store is when it retains low granularity of data over long periods of time. System performance is affected when performing analysis on low-granularity data because the operational data store becomes large and cumbersome.

Analysis is done by querying (using structured query language or off-the-shelf query tools) and reporting, without impacting the performance of the production systems. The drawback of the operational data store is that because it is not designed for decision support applications, complex queries may result in long response times and heavy impact on the transactional systems (Murtaza, 1998).

The *decision support data warehouse* architecture simply consists of snapshots of corporate information with low-level or highly summarized data (Figure 5.1) (Murtaza, 1998). Advantages of this method include minimal infrastructure costs, access to nonvolatile data, quick deployment time, and no repetitive data stores. But the main disadvantage with this architecture is its inherent lack of flexibility to handle complex decision support analysis. The decision support data warehouse provides good historical data but fails to optimize access to the data.

The *data mart* appears to be the most common data warehouse application today. Data marts store nonvolatile, time-variant, and summarized information used to serve the information needs of the business unit (Murtaza, 1998). It differs from a data warehouse in that a data mart contains customized data to support particular analysis requirements (Moody and Kortink, 2000), whereas the data warehouse data is truly corporate (Inmon, 1996a). Peacock (1998) further explains

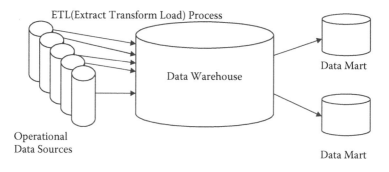

Figure 5.1 Data warehouse architecture.

the difference between a data warehouse and a data mart. A data warehouse is an enterprise-level data repository that draws its contents from all critical operational systems and selected external data sources. The data warehouse is based on a data model that is a time- and resource-consuming cross-functional effort. A data mart, in contrast, is a functional subject area or departmental data repository that draws its contents from systems that are critical to the department and from selected external sources.

O'Sullivan (1996) points out that the scope of the system determines whether it is a data warehouse (enterprise-wide) or a data mart (departmental or functional). Depending on the architecture chosen, a data mart can be constructed as individual components within the scope of a comprehensive data warehouse plan (top-down design) or due to cost and time restraints, built independently of data warehouse initiatives (bottom-up design) (see Figure 5.2). A major drawback of the bottom-up approach, as pointed out by Peacock (1998), is that the architecture of the data mart may be inconsistent with the architecture of other data marts and the data warehouse when it is finally built. Proliferation of independent data marts can yield fragmented, inconsistent data that could inhibit future development of cross-functional information. But the advantage of a data mart over a data warehouse is that it offers lower entry costs and faster implementation than a data warehouse, which typically involves a data modeling effort encompassing enterprise-wide information requirements (Sigal, 1998). Hoven (1998) points out that a data mart can be a practical first step to gain experience toward building and managing a data warehouse. A data mart could be built as a subset of a data warehouse (Moody and Kortink, 2000), focusing on delivering value to a specific business area. With proper planning, multiple data marts may be gradually consolidated under a common management umbrella to create an enterprise data warehouse.

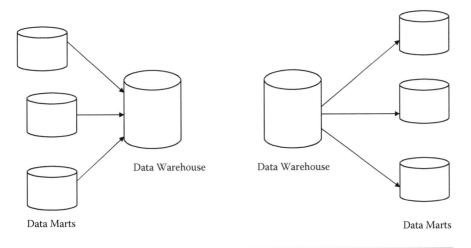

Data Warehouse   Data Warehouse

Data Marts   Data Marts

Figure 5.2   Data warehouse/data mart architecture.

Depending on the data warehouse environments, different data warehouse architecture designs are developed to suit the particular situation. A popular architecture design for quick prototyping of the data warehouse follows a MOLAP- or ROLAP-centric (multidimensional or relational OLAP) route. Data from one or more operational systems for a subject area is loaded into a MOLAP or a relational database. As users grow familiar with the online analytical processing, the user community widens. Users feel the need for additional subject areas and more data marts are built to satisfy those demands. This in turn leads to the need for a more integrated approach to load the data marts. This leads to the development of a global repository where data is cleaned uniformly and derivation rules for data transformation are standardized. The repository acts as an intermediate stage between the operational source systems and the data marts. The intermediate repository usually has low granularity and rich dimensionality. Data marts are developed based on the intermediate repository to enhance user interfaces and to provide better performance and richer functionality. The building of a repository retrospectively is usually more expensive than developing it earlier.

The quick-delivery-of-results approach has certain disadvantages. Because no underlying architecture is built to support the prototype, a collection of MOLAP or ROLAP prototypes over time is mistaken for a data warehouse. The lack of an integrated architecture adversely affects future needs for enhanced or wider analytical information. An integrated architectural design facilitates future needs for other analytical information.

Even though organizations can select data warehouse architectures that support their strategic vision, it has been seen that implementation efforts often fail (Vatanasombut and Gray, 1999). Implementation efforts may fail due to difficulty in use (Guimaraes et al., 2003), lack of support from management (Hwang et al., 2004), or inability to respond to business changes (Armstrong, 1997). The next chapter discusses factors that affect the implementation and adoption of the data warehouse in organizations.

Chapter 6

Factors Influencing the Success of a Data Warehouse

Success of an information system has been measured by researchers in various ways, including user satisfaction (L. Chen et al., 2000; DeLone and McLean, 1992), data quality (Ballou and Tayi, 1999; McFadden, 1996), return on investment (Cooper et al., 2000; B. Shin, 2003), and perceived benefits (Ballou and Tayi, 1999; DeLone and McLean, 1992). Weir et al. (2003) and Watson and Haley (1998) have noted that it is difficult to put a financial value on an intangible benefit such as data or information. Wixom and Watson (2001) point out that data warehouses have unique characteristics that may shift the importance of factors that apply to it. Schubart and Einbinder (2000) have focused on future usage and perceived effectiveness as measures of success for a data warehouse. Long-term success of the data warehouse depends on the organization's ability to use the data warehouse to fulfill its strategic milestones (Weir et al., 2003).

According to researchers (Hwang et al., 2004; Wen et al., 1997), data warehouse projects have a high possibility of failure. Wixom and Watson (2001) estimate that one half to two thirds of all data warehousing fails. Although there is a fundamental change in the business environment, with demands on gathering new data, new levels of data integrity, and consistency, just building data warehouses is not the solution. An organization needs to thoroughly understand the impact of a data warehouse on its operation before writing a single line of data warehouse code. For example, in a banking environment one needs to know exactly what kind of data will be required for compliance and reporting, for measuring performance, and for calculating compensation before embarking on a data warehouse project (Dembo, 2004).

Companies are integrating their data and building data warehouses to create advantages, send new products and services to markets faster, provide improved customer service, and reduce production and inventory costs. However, many firms are failing to realize these benefits (Johnson, 2004). For some, data warehouses created to integrate data from multiple sources have a user interface that is difficult to navigate (Watson and Haley, 1998), or else there are misunderstandings about expected service levels. For others, the generated data turns out to be inaccurate or irrelevant to the users' needs (Ballou and Tayi, 1999; Strong et al., 1997), or delivered too late to be useful. Thus, it is essential to understand the factors that ensure a successful adoption of data warehouse technology. Most of the researchers in the literature review focus on the technological and operational aspects. Very little research has considered factors at the managerial or strategic levels. This research seeks to address this weakness.

This chapter identifies and discusses the factors that influence the success or failure of a data warehouse. Numerous success factors affect the implementation of data warehouses. These factors are grouped into four broad categories to facilitate a comprehensive discussion: (1) organization factors, (2) user factors, (3) technology factors, and (4) data factors.

Organization Factors that Influence Success of a Data Warehouse

In the organizational dimension, factors such as the size of the organization, top management support, existence of a champion, team skills, organizational barriers, and organizational culture can all affect the adoption of a data warehouse technology (Figure 6.1).

Hwang et al. (2004) point out a strong correlation between the degree of business competition and the adoption of new information technology. Enterprises try to raise their perceived competitive advantage by adopting new technology, especially if their competitors have adopted or are adopting this new technology. Referencing a study of the banking industry in Taiwan, Hwang et al. (2004) conclude that the larger the *size of the organization*, the more resources and capital can be allocated to adopt new information technology.

The greatest potential benefits from data warehousing come when the data warehouse is used in the redesign of business processes and to support strategic business objectives (Watson and Haley, 1998). Securing top-level management's commitment and support is essential before embarking on a data warehouse project (Figure 6.2). *Support and commitment from the top management* is considered

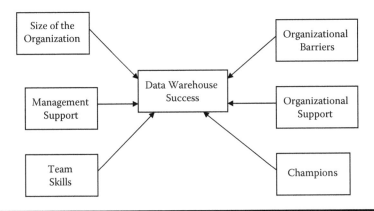

Figure 6.1   Organizational factors influencing data warehouse success.

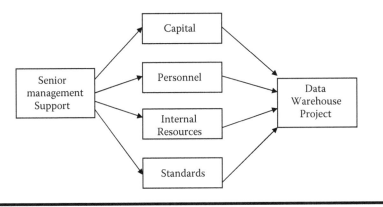

Figure 6.2   Management support.

important to secure the required capital, human support, and internal resources during the adoption and development processes (Chenoweth et al., 2006; Hwang et al., 2004). Because standardization of information is a key rationale for a warehouse, inadequate coordination of participants can be its downfall (O'Sullivan, 1996). A strong mandate from senior management is needed to impose standards, because different areas within a company may resist changing their ways.

In many companies the decision-making environment reflects a fractured approach. In such organizations different departments create separate small databases to support disjointed operational systems (Oates, 1998). Vital information exists in dozens of separate and unrelated databases, making an integrated view of the business impossible (Hurd, 2003). An enterprise-wide approach to information management in the form of a data warehouse can address this problem to some extent by integrating these islands of data (I.J. Chen and Popovich, 2003).

However, information silos often exist in organizations for reasons that have nothing to do with information technology. Individual departments many times have little control over the budget for an enterprise-wide approach to information management, but they have a great deal of control over the budget for their own silo within the company. Breaking down these *organizational barriers* is often more difficult than surmounting the technical ones. Hurd (2003) suggests that strong leadership is required to drive individuals within an organization to overcome traditional constraints and work together. *Champions* are important to data warehousing as well as other IT projects because they actively promote the project and provide information, material resources, and political support (Counihan et al., 2002; Hwang et al., 2004; Sammon and Finnegan, 2000; Watson et al., 2002; Wixom and Watson, 2001). A champion among senior management is especially important in an enterprise-wide data warehouse where the warehouse builds on a number of subject areas across different business units, because many of the benefits are intangible and not always foreseeable. Apart from senior management commitment, data warehousing projects also need business and IT commitment (Figure 6.3).

In the project planning dimension, the *skills of the data warehousing development team* can have a major influence on the outcome of the project (Cooper et al., 2000; Hwang et al., 2004; Wixom and Watson, 2001). A highly skilled project team is better equipped to manage and solve technical problems. Hwang et al. (2004) point out that the selection and inclusion of appropriate users in the project team is also important. The skills of the data warehouse teams have an effect on end-user participation, which in turn has a direct impact on the adoption of data warehouse technology (Wixom and Watson, 2001). User participation is essential for better communication and coordination of the users' needs because this ensures the system's successful implementation. End-user participation helps the development team manage users' expectations and satisfy user requirements.

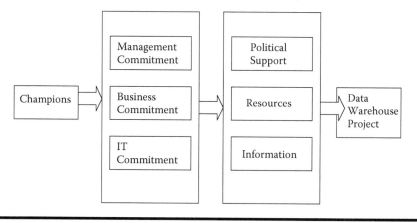

Figure 6.3 Champions.

A data warehouse requires detailed *planning* involving internal operational procedures and workflow reconfiguration (Ang and Teo, 2000). A data warehouse implementation is a major event and is likely to cause organizational perturbations. Management issues pertaining to the data warehouse project provide important considerations for the project team. In the project planning dimension, coordination of organizational resources including money, people, and time ensures completion of the project and ultimately affects the adoption of data warehouse technology.

Cultural factors affect the planning, implementation, and operation of IT applications. In organizational cultures that are data driven and data accountable, users will demand the richness of information housed in a data warehouse (Schubart and Einbinder, 2000). *Organizational culture* consists of the shared assumptions, beliefs, and values that exist within an organization and how the behavior of the people is influenced by it (Doherty and Doig, 2003; Pliskin et al., 1993). With increasing integration of the global economy, one cannot ignore the cultural issues. Shanks and Corbitt (1999) suggest that culture plays a significant role in information systems during requirements gathering and in the construction of quality practice. Many IT projects fail due to a poor fit with the prevailing culture or a failure to build a culture to support change (Pliskin et al., 1993). According to Pliskin et al. (1993), failure may occur when there is a clash between the cultural presumptions embedded in the system design and the actual culture of the implementing organization.

Data warehousing systems have to be organization specific to meet the specific needs of the organization (Figure 6.4). Every organization represents a specific organizational structure, architecture, as well as political and financial constraints. A data warehouse project has to encounter organization-specific issues. The nature of the business of the organization, the business environment within which the organization operates, the technical sophistication of the organization, as well as its information intensity affect the data warehouse project. The strategic objectives of

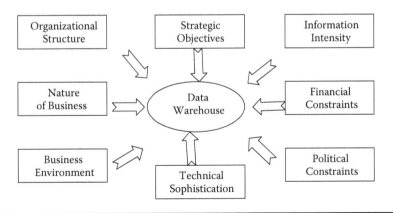

Figure 6.4   Organization-specific issues affecting data warehouse.

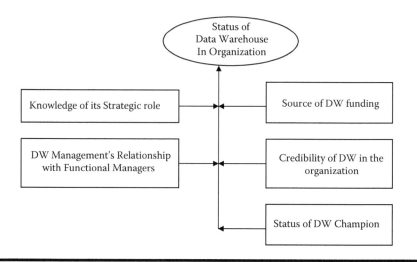

Figure 6.5 Status of data warehouse.

the organization affect the scope of the data warehousing project and the strategic uses of the data warehouse.

Data warehousing is usually considered a secondary activity in an organization even though it can offer great potential for the organization's business. A data warehouse can support and transform an organization's business, as well as enable business initiatives, but its success is often judged in terms of speed, quality, and cost. A data warehouse can play a strategic role in an organization and can have a high impact. Data warehouses can help leverage information, realize business opportunities, and provide information readiness to respond decisively. Yet it usually has a low status. Status concerns rank and importance. Status provides the power to influence others in the organization and control the resources. (See Figure 6.5.)

Status has no quantitative indicators and its assessment is largely subjective. It can be assessed through observation and an analysis of organizational culture. The position of the data warehouse management head within the organization as well as the resources available to it can reflect the status of the warehouse within the organization. Higher management must be involved in an enterprise-wide data warehouse initiative due to the large expenditure involved. If higher management is involved in data warehousing efforts it should be seen as having higher status.

One reason data warehousing often has lower status in the organization is due to the lack of knowledge of its role by others in the organization. The data warehouse is seen as a supporting tool for decision making and therefore is not viewed as a strategic tool. When a data warehouse holds a lower status in the organization, it leads to difficulty in negotiation and manipulation for resources on its behalf and it also faces user resistance. On the other hand, if a data warehouse is viewed as strategic for an organization, the data warehouse management is likely to have increased power.

The status of the data warehouse is also affected by its management's relationships with other functional managers and users within the organization. The issue of credibility is important for a data warehouse. When a data warehouse is perceived as having credibility within the organization, it can be used to effect changes within the organization. A strategic alliance between different functions or departments within an organization and the data warehouse management ensures a more successful data warehouse. A data warehouse can bring about changes in an organization. However, to bring about changes, a data warehouse must have sufficient credibility and status in the organization.

The way the data warehouse project is funded or justified might be indicative of its status within the organization. A poor implementation or a poorly understood data warehouse reduces its impact on the organization. If the data warehouse team has difficulty communicating its success, others in the organization will not perceive the data warehouse as being successful.

A powerful champion of the data warehouse may increase its status. A high-status champion may influence others and take an active part in decision making in the organization. The data warehouse objectives need to be closely aligned with the strategic objectives of the organization. The information in the data warehouse would then be seen as valuable and the data warehouse will not be undervalued.

On the other hand, technology and information systems have had a marked impact on the way work is organized, allotted, and accomplished in modern organizations (Cooper et al., 2000). Doherty and Doig (2003) suggest that implementation of data warehouses can have a significant impact on the host organization's culture. Major changes and improvements to the flow and quality of information may have the potential to modify organizational culture, particularly in the areas of customer service, flexibility, integration, and empowerment. Doherty and Doig (2003), in their study of data warehouse implementations, found that to realize these benefits, changes were required in working practices and employee behavior, which could in turn have cultural implications.

The success of a data warehousing project does not rest entirely on technology. An organization may build a data warehouse with the right tables, middleware, and technology structure to support it, yet the data warehouse may fail. This is because delivering value to the users and maintaining the data warehouse are of equal importance. The culture of the organization has to move from a transaction processing to a decision support mindset. With flattening of the organization structures from purely hierarchical structures to ones where decision making is more distributed, information and data need to be delivered to a wider set of end users. A data warehouse should be able to deliver decision support to this wider base of users.

User Factors that Influence Success of a Data Warehouse

Data warehouses provide decision support to organizations with the help of analytical databases and online analytical processing (OLAP) tools. A data warehouse by itself does not create value (Watson and Haley, 1998). Value comes from the use of the data in the data warehouse. One of the important determinants of new technology acceptance is the perceived ease of use and perceived usefulness (Guimaraes et al., 2003; Nah et al., 2004). It has been observed (Gorla, 2003) that despite the potential benefits of data warehousing, corporations often do not provide tools to end users that they can use easily, resulting in users not utilizing the tools, millions of dollars of unused software, and unrealized return on investment. (See Figure 6.6.)

Nah et al. (2004), in their investigation of *end users' acceptance* of enterprise systems, found that factors such as perceived compatibility, perceived ease of use, and attitude were significant determinants in the adoption of a system. They found that in order to create positive acceptance among end users, organizational interventions should focus on the issue of compatibility as well as the issue of technology fit with organizational context. For a system to be accepted by its end users, not only must it be perceived as useful and easy to use, but it is also important that the end users perceive the system to be compatible with their values and past experiences and to be a good fit with the organizational context (Figure 6.7).

The success of a data warehouse depends heavily on *end-user satisfaction* (Ang and Teo, 2000; L. Chen et al., 2000; DeLone and McLean, 1992; McFadden, 1996; Wixom and Watson, 2001). Developing a data warehouse is a difficult endeavor, but realizing significant benefits is much more difficult (Ang and Teo, 2000). As such, users must undergo continual, formal, and systematic training to get the most from the data warehouse. Technical system quality is important to the success of a data warehouse, but just as important is the need to understand and address the human issues involved.

The importance of user-related factors such as *user participation*, user training, and user acceptance in the success of a system is also recognized by Guimaraes

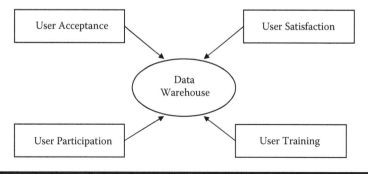

Figure 6.6   User-related factors affecting data warehouse adoption.

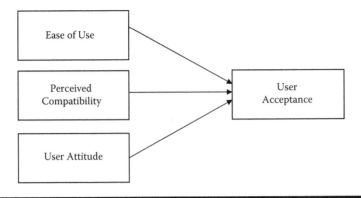

Figure 6.7   User acceptance.

et al. (2003). They note the importance of user training as a significant factor for user participation and promote user–developer communication during the system development process to reduce user conflict.

Apart from user satisfaction, the growth of the user base is perceived as a key indicator of the success or failure of a data warehouse (Armstrong, 1997). If the data warehouse is seen as providing timely access to valuable information, then the user base grows rapidly. On the other hand, if the system is perceived as low in information value, difficult to use, or lacking in capability, there will be no growth in the user community. Growth in the number of users could be the actual user community, the number of users logged on to the system, or the number of concurrently active users.

Technology Factors that Influence Success of a Data Warehouse

Other reasons for failure of data warehousing projects have been the *complexity of large-scale data warehousing* and the conflicts arising from data mart activities of business units that cannot resolve conflicting objectives. To address these problems, Sigal (1998) suggests a hybrid deployment strategy. Due to business objectives and cost, time, and skills constraints, he advocates a faster bottom-up data mart deployment strategy combined with a top-down high-level data model. Such a hybrid deployment approach begins with the development of one or more data marts on a staged basis. Capabilities of cross-functional processing are anticipated during the planning phase and are implemented incrementally as the system grows.

Prototyping the data warehouse as part of the implementation process is an important factor in the success of a data warehouse. A typical prototyping practice in a data warehouse implementation involves the construction of a small data mart for an important area for which the data warehouse is being implemented.

A prototype supports better understanding and functioning in an important area within the organization and helps obtain buy-in on data warehouse usefulness by organizational personnel (Little and Gibson, 2003). Tyagi (2003) adds that implementing a well-designed pilot helps in anticipating costs and identifying potential stumbling blocks. Using a phased-in implementation increases the chance of success because it enables managers to monitor data integrity and system quality issues step by step. Companies also need to avoid "scope creep" once a warehouse project has been implemented (L.K. Johnson, 2004).

Standardization in the technology platform can realize several benefits as well. Having fewer different technologies will result in faster and less costly implementations, because there will be fewer interfaces and incompatibilities. The number of project failures will be reduced because of better familiarity with the technology and its capabilities. Time spent evaluating, selecting, and learning new technologies would also be saved (Sigal, 1998).

A basic requirement for a data warehouse is the ability to provide users with *accurate* (Cui and Widom, 2003; Rahm and Do, 2000) and *timely* (Inmon, 1996a; Kimball, 1996; Schubart and Einbinder, 2000; Squire, 1995) *consolidated* (Chaudhuri and Dayal, 1997; Golfarelli and Rizzi, 1998; Moody and Kortink, 2000) information as well as a fast query response time (Bernardino and Madeira, 2000; Datta et al., 1998). For this purpose, stored result sets or materialized views are used so that a query is answered more quickly against the materialized view than querying directly the base data stores (Mistry et al., 2001; Yang et al., 1997). A materialized view is a snapshot or replica of a target master from a single point in time. These materialized views need to be updated consistently (Gupta and Mumick, 1995). In contemporary organizations this becomes an important issue as more firms are engaging in business-to-business (B2B) commerce and data exchange for seamless decision making across the value chain (Triantafillakis et al., 2004). Boundaries of organizations have become more fluid and the data sources are no longer entirely internal. In these extended enterprises, data from outside the organization has to be integrated into a single repository.

One key stumbling block to rapid development of data warehouses is *warehouse population* (J. Srivastava and Chen, 1999). The general conclusion is that it is labor intensive, error prone, and generally frustrating, leading to a number of data warehouse projects being abandoned midway through development. Problems arise in integrating the data due to semantic discrepancies, scalability issues, and incremental updates. Semantic discrepancies arise when integrating data instances from multiple sources. Semantic discrepancies could manifest as problems in entity identification or attribute value conflicts (J. Srivastava and Chen, 1999). Scalability tasks can be very complex and encompass the capacity to store immense data, the ability to efficiently process queries, the capability to perform data management operations, and delivering business-critical availability, all at huge scale (Shahzad, 1999). Problems with incremental updates arise when data is added to an existing warehouse and must be integrated with preexisting data (J. Srivastava and Chen, 1999).

The growth of the Internet has changed the way information is managed and accessed today. The availability of commercial data on the Web has given rise to the need to analyze and manipulate these data to support corporate decision making (Bhowmick et al., 2003a; Bhowmick et al., 2003b; Kosala and Blockeel, 2000). The convergence of data warehousing and the World Wide Web leads to the rise of *Web warehousing* (Bhowmick et al., 2004; Bhowmick et al., 2003b; Madria et al., 2003; Zaiane et al., 1998). A Web warehouse delivers the same kinds of applications that a data warehouse solution delivers but via the Web technology as opposed to the client server technology (Tan et al., 2003). The Web technology adds the capability to perform search, statistical analysis, and mining operations with business-related non-data objects made up of pictures, sound graphics, video, and more. Web-based technologies are rapidly becoming a part of the business fabric (N. Jones and Kochtanek, 2002). With the rapid evolution of the Internet and the growth of B2B e-commerce (Zeng et al., 2003b), many organizations will look to new and innovative data mining and data warehousing technologies, such as virtual data warehousing, to meet the increasing demands (Genesereth et al., 1997).

As business needs change over time, a data warehouse needs to be responsive to these changes to be successful. A data warehouse provides customers with information to run their business. If the data warehouse cannot adapt to changes in the environment, then the company loses the advantage that the information provides (Armstrong, 1997). A warehouse needs to be built with a solid foundation that is *flexible* (Rundensteiner et al., 2000; Wixom and Watson, 2001) and *responsive to business change* (Armstrong, 1997). According to Armstrong (1997), this concerns three main areas: (1) the database, (2) the application middleware, and (3) tool integration. For the data warehouse to have long-term success, all three areas must have scalability, high availability, and robust manageability. Change management (Bliujute et al., 1998) would span all the components of a data warehouse and would play a vital role in the development and overall success of a data warehouse. To successfully manage a data warehouse, Sen (2004) argues that two conflicting goals need to be managed: maximizing the use of the data warehouse asset while consistently achieving user expectations by continuously monitoring the effect of business change.

Data Factors that Influence Success of a Data Warehouse

The data warehouse is an environment for the collection, management, and distribution of data from various sources to the end users (Armstrong, 1997). It allows users to ask questions across cross-functional data and gain insight when the need arises. Nevertheless, many data consolidation projects underperform because of bad data. As data warehouses evolve, myriad *data quality* issues emerge in the form of repetition, different formats, different metrics, and multiple business definitions (Sinn, 2003).

Data Quality

The data warehouse contributes to the decision-making power of its users. The quality of these decisions depends on the quality of the data used to arrive at these decisions. In a data warehouse, data is consolidated from multiple sources and integrated into a repository for easier access and analysis. The concern for quality of the data in the data warehouse is high because poor data quality is one of the leading reasons for the failure of data warehousing efforts. Data quality problems are often regarded as being caused by errors in the data. This is an intrinsic view of data quality. Problems in data can arise due to larger problems in the way the data was defined for use in the data warehouse or the way it was produced. A data warehousing effort may suffer because it does not store the appropriate data required by its users. In the data warehouse, data quality issues have to be considered in light of the consumers who use it.

In the data warehouse, there are three main stakeholders of data: the data producers, the data custodians, and the data consumers (Figure 6.8). Data is produced in multiple source systems that are then transformed into useful information by the data warehouse users who are the data consumers. The data custodians provide resources for collecting, storing, and maintaining the data for use by the data consumers. High-quality data is data that is useful to its consumers.

Because data is produced in the source systems for multiple purposes that may be different from its purpose in the data warehouse, using the same data for purposes other than its original purpose can lead to quality issues with the data. The issues of data quality in the data warehouse can be characterized into four main categories: intrinsic, accessibility, contextual, and representational (Figure 6.9).

Problems in any of these areas affect the high quality of data in the data warehouse. Addressing data quality problems in a data warehouse involves identifying the data quality problem, determining the causes, and resolving the data problem. This may require data analysis, process improvements in producing, transforming,

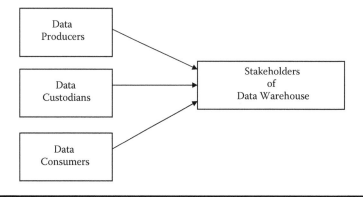

Figure 6.8 Stakeholders of data warehouse.

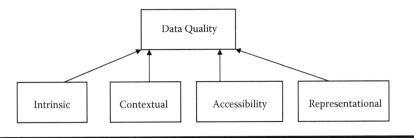

Figure 6.9 Data quality.

and using the data, and cleaning the data itself. Because warehoused data can support multiple decision processes, the data quality enhancement is further complicated when different users require different levels or characteristics of quality. Furthermore, because data warehouse users assess data quality in context to their ever-changing requirements, enhancing data quality is an ever-moving target.

Intrinsic Data Quality

Data is integrated into the data warehouse from multiple source or legacy systems as well as some sources external to the organization. Similar data may be captured in these source systems that may appear as conflicting in the data warehouse. These mismatches in data may be due to differing levels of accuracy of data in the source systems, leading to questionable data in the data warehouse from the user's perspective. This leads to believability concerns of the data warehouse data, making users less likely to use it. This affects the value the data warehouse data brings to the decision-making process of its users. (See Figure 6.10.)

Intrinsic data quality problems also arise when data production is subjective or requires human judgment. Interpreted data is of lower quality than uninterpreted raw data. For example, data abstracted from a doctor's report by data abstractors

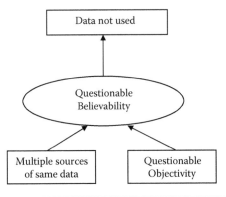

Figure 6.10 Intrinsic data quality.

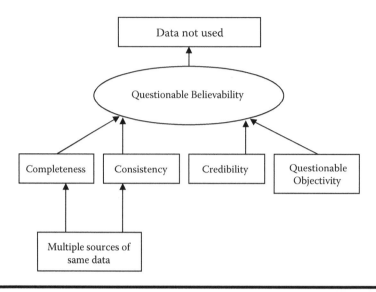

Figure 6.11   Intrinsic data quality problems.

is subject to the knowledge, experience, training, and understanding of individual data abstractors. Believability problems also arise when information is codified into data. Problems with subjective data production often lead to concerns of data objectivity over time and reduced use of data. (See Figure 6.11.)

Contextual Data Quality

Data in the data warehouse is produced at multiple legacy systems and used by multiple users with differing data quality needs. Users define high-quality data based on their requirements, and these may differ among users for the same data and even for the same user over time as his or her requirements change. High data quality is maintained by devising flexible approaches to providing data so that data is useful and valuable to the users in the context of their needs. Data that can be manipulated easily to meet changing requirements is considered to be of higher quality.

Incomplete data, missing data, inappropriate data, and data that is not adequately defined cause contextual data quality problems. Incomplete data may arise due to operational problems in the design of the transactional system, problems with capturing the relevant data, or problems with the amount of data.

Timeliness of data is another attribute that affects contextual data quality. Issues with timeliness of data in the data warehouse may be due to the time it takes to update data in the warehouse, or it may be due to the manner in which data is communicated. For example, large volumes of data can lead to delays and timeliness issues in data warehouses. There are usually tradeoffs involved in improving

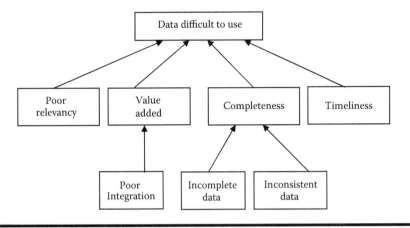

Figure 6.12   Contextual data quality.

data quality based on limited resources. For example, data may be made available, but at the cost of completeness or accuracy. (See Figure 6.12.)

Accessibility

Failure to store appropriate data is sometimes viewed by users as a quality problem. Nonavailability of external data that might be necessary for decision making can also be viewed as an accessibility problem. Another accessibility problem with data quality occurs when required data is represented in images, coded, or text form. This data is not easily accessible and creates a data quality issue because it is difficult to analyze and interpret. For example, it is difficult to analyze trends from images of X-rays across patients in a hospital. Access security can also cause data quality problems. In a healthcare or research environment, permission to access the patient data based on identifying information is usually restricted. Patient confidentiality concerns limit data access to users with prior approvals. This barrier to open access is sometimes viewed as a data accessibility problem. (See Figure 6.13.)

Representational Data Quality

Inconsistent definitions across different systems within an organization and inconsistent data representation across systems gives rise to representational quality issues. For example, if the data element "race" or "ethnicity" has different definitions across different healthcare systems, it is difficult to use the available data for research on quality of patient care. Again, if the measure for a data item is different across systems, it is difficult to interpret the data easily. Ease of understanding, ease of interpreting, and consistent representation of data are necessary to enhance data quality.

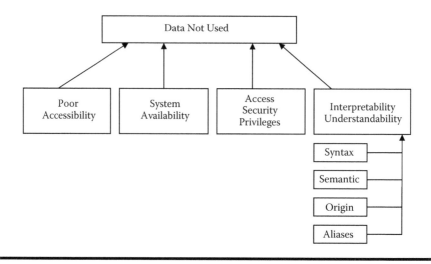

Figure 6.13 Accessibility problems affecting data quality.

Data quality assurance must be integrated into the data warehousing process from planning to implementation to ensure good data quality that supports organizational activities. Because data warehouse users access quality relative to their requirements and needs, enhancing data quality in the data warehouse environment means addressing a variety of issues. Data quality could be enhanced by addressing the issues in a systematic manner. To improve data quality, the organizational activities the data warehouse supports needs to be determined. Next, the data sets that support the organizational activities need to be identified. The quality of the data sets along data quality dimensions needs to be identified. This would include identifying both existing and potential problems with the data sets. Projects would then have to be identified to be undertaken to enhance data quality. The cost and effect of the project on the anticipated data quality of the various data sets would have to be estimated. In reality, the difficulty lies in the implementation of the project, due to the many tradeoffs involved. Beginning data quality enhancement efforts with activities with the highest priorities would have the most impact.

In data warehouse projects, data quality is an ongoing concern. According to Redman (1995), errors in data can cost a company millions of dollars, alienate customers, and make implementing new strategies difficult. He proposes a three-step strategy to improve data quality: (1) identifying data quality problems, (2) treating data as an asset, and (3) applying quality systems to the process that create data.

Sinn (2003) suggests a *data stewardship program* as a solution to ensure good quality data. The main function of a data stewardship program is the management of an organization's data assets in order to improve and maximize the data's accessibility, reusability, and quality. Data stewardship is accomplished by ensuring the engagement of senior management and the building of cross-functional teams comprising business and technical sides. Cross-functional teams ensure that everyone understands their role in maintaining the quality of data. Increasing their awareness of what data exists and where they can find it makes it more likely they will actively use the data. Moreover, the technical staff gains insights that allow them to fully align their work priorities with the company's overall business strategy. Similarly, Armstrong (1997) points out that the issue of data quality must be viewed from a business-wide sense, because it is only when the data is compared to other elements that the quality is known.

Evaluation is another way to deal with poor data quality. Examining overall data quality, performing gap analysis to identify areas that lead to bad data, and reviewing existing data management programs can all help toward improving data quality (Sinn, 2003). L.K. Johnson (2004) lists eight metrics to evaluate data warehouse performance: (1) the percentage of data tables completed; (2) connectivity; (3) data integrity; (4) the usage rate of data tables and reports; (5) the delivery time for requested data tables; (6) the number of days the IT group needs to resolve problems; (7) the number of times per day the users "hit" the data warehouses' databases; and (8) the run time.

Despite various attempts, data quality problems continue to persist in data warehouses. Poor-quality data enters from operational databases and other sources of data into the data warehouse. The convention is to apply *data integrity* practices as a one-time process when the data enters the database. The failure to link integrity rules to organizational changes is one of the reasons data quality problems persist. One mechanism to solve this problem is to embed data integrity in a continuous data quality improvement plan. Y.W. Lee et al. (2004) suggest an iterative data quality improvement process as data integrity rules are defined, violations of these rules are measured and analyzed, and then these rules are redefined to reflect the dynamic and global context of business process changes.

The effectiveness of decision making is influenced by various factors such as time constraints, time pressures, and data quality. Experienced decision makers who have worked in a particular milieu for a sufficient period of time develop a feel for the nuances and eccentricities of the data used and intuitively compensate for them. As organizations move to stored repositories such as data warehouses, this intuitive feel is not preserved for many who extract data from it. As users are increasingly removed from any personal experience with data, knowledge that would be beneficial in judging the appropriateness of the data for the decision to be made is lost (C.W. Fisher et al., 2003). *Data tagging* is a way to capture some of the knowledge regarding the data's quality and origin along with the actual data values. This is accomplished by incorporating technical metadata directly into the data warehouse design and architecture. These technical metadata tags are referenced at a low level of granularity in the data warehouse. Data quality information (DQI) is metadata that can be included with data to provide the user information regarding the quality of that data. C.W. Fisher et al. (2003) find that it is expensive in general to generate and maintain such information. Their overall conclusion is that DQI should be made available to managers without domain-specific experience. It should only be incorporated into data warehouses used on an ad hoc basis by managers.

Chapter 7

Strategic Alignment

The data warehouse has emerged as a powerful tool in delivering information to users, creating competitive advantage, and building support for decision making and customer satisfaction. Data warehouses have unique features that make them different from other decision-support applications. Data warehouses also differ from traditional operational systems. The data warehouse implementation process has an enterprise-wide impact on the infrastructure of the organization. Factors that affect data warehouse implementation applying IT implementation knowledge have been investigated and various success factors for data warehouse implementation have been identified in Chapter 6. These include organization factors (project sponsorship, champions, management commitment, team skills, organizational culture), user factors (user acceptance, user participation, perceived benefits), technology factors (availability of resources, complexity, architecture selection, standardization, consolidation, warehouse population, flexibility), and data factors (data quality, evaluation, data integrity). But data warehouse implementations are still known to fail. This is because each data warehouse system has an organization-specific set of requirements, constraints, issues, and implications that need to be addressed. The data warehouse needs to be linked to the organization's objectives.

A data warehouse is built to address the strategic needs of an organization. Organizations allocate considerable resources to data warehouse projects, but there has been very little discussion on how to achieve a strategic alignment between the data warehouse and the business needs, to ensure its success. Discussion on managerial or strategic issues of data warehousing have been rare. Although the need for commitment and support from top management has been identified as a critical factor, no specific guidelines have been proposed on how to attain this.

This chapter examines the theoretical concepts and current research on strategic alignment with respect to IT. This chapter provides a theoretical grounding to explore and examine how strategic alignment between the data warehouse and the business lends itself to a successful implementation of the data warehouse, which is discussed in Chapter 8.

This chapter is divided into five sections. The first section discusses the need to align IT with business strategy. The second section presents the strategic alignment model and its constituents. The third section discusses recent developments in strategic alignment research. The fourth section presents the alternatives to the strategic alignment model. The fifth section presents the factors that enable alignment between IT and business.

Need for Strategic Alignment Between IT and Business

The use of information technology in business has transformed over the last several decades from operational utility in the 1960s to that of a competitive weapon today (Bakos and Treacy, 1986; Pollalis, 2003). This phenomenon has affected the ways organizations are managed (Brynjolfsson and Hitt, 2000) as well as the way IT affects the strategic activities of an organization (Brynjolfsson and Hitt, 2000; Loebbecke and Wareham, 2003; Luftman et al., 1993). Technology is enabling and causing changes that are so substantive and persuasive that it is no longer possible to have a disconnect between an organization's strategic plans, goals, and directions and its IT initiatives, resources, and management (Hitt, 2001).

The prevailing hyper-competitive markets (Kandampully and Duddy, 1999) bring pressure for businesses to shorten product life cycles (Griffin, 1997; H.L. Lee, 2002; H.L. Lee and Whang, 2001; Calantone and Di Benedetto, 2000), quickly identify and penetrate new market segments (Kaplan and Norton, 1992; R.K. Srivastava et al., 1998), increase operational efficiencies (Kärkkäinen and Holmström, 2002; Vander Vennet, 2002; Sarkis, 2000), and disintermediate supply chains and distribution channels (Rabinovich et al., 2003; Tillquist, 2002; Ho et al., 2003). Businesses seek sustainable competitive advantage in these markets by leveraging technology to the fullest extent. In these markets, alignment between the business strategy and information technology is not a luxury but is a cost of entry (Bruce, 1998).

The number of technologies and software capabilities that exist today are more than what a business could ever possibly adopt. The key issue for companies is not the availability of technology but choosing which technology to deploy and to what purpose. Businesses invest billions of dollars in information technology, yet studies like Ryan and Harrison's (2000) indicate that more than 50 percent of IT implementations actually cost more than twice their original estimates. A lack of foresight in the IT investment decision process has been cited for this diminishing payoff (Schniederjans and Hamaker, 2003). Others cite a need to deploy information technologies in ways that are of the most relevance to the businesses and their strategic objectives (Andal-Ancion et al., 2003; Kearns and Lederer, 2003; Tallon et al., 2000).

The concept of strategic alignment is not new; it is more than two decades old (McLean and Soden, 1977; Chan, 1996a,b; Henderson and Venkatraman, 1990). But improving IS/IT strategic planning continues to rank among the major issues facing IT executives. Strategic alignment has been identified as one of the most critical IS research issues facing academic researchers (G.G. Lee and Bai, 2003). IS strategic alignment has been among the top five challenges faced by senior executives over the last decade (Chan, 1996) and continues to be of increasing importance today. The reason for the interest in strategic IS alignment is because it has shown to enhance not only IS success but organizational success as well (Hirschheim and Sabherwal, 2001). Despite the recognition of the importance of strategic IS alignment, insufficient research has been conducted on how to achieve and sustain it. The difficulties in achieving and sustaining alignment have been underestimated. The path toward alignment is not an easy one.

As Henderson and Venkatraman (1989) note, strategic IT planning has evolved over the last three decades. IT planning first focused on effective allocation of the firm's resources to IT (Teo and Ang, 1999; Lederer and Mendelow, 1988). IT planning focused on the automation of processes (Mukhopadhyay et al., 1997; Bresnahan et al., 2002). The planning process employed a functional model of the business as a frame of reference. The IS planning product was a set of functional applications like marketing systems (Glazer, 1991) and financial systems (Krishnan et al., 1999; Brynjolfsson and Hitt, 1995). IT architecture was developed and implemented through a series of segmented projects.

In the next era, the enterprise was the context of IS planning (Davenport, 1998; Umble et al., 2003), and cross-functional integration (Fiedler et al., 1994) became the primary focus. IT planning decisions were extended beyond the project level. It called for an integrated strategy that explicitly recognized the potential for IT to enable business strategy.

Henderson and Venkatraman (1989) note that IT strategy is now in a third era, in which business strategy is not viewed as stable. The organization is viewed as facing a dynamic environment. The information technology market at the same time is also dynamic. IT is viewed as an opportunity to enhance the competitive capability of the firm (Boynton and Zmud, 1987; Ives and Learmonth, 1984). The IT planning process now must not only leverage emerging technology, which is vital to business strategy, but also organize the delivery of IT products and services to meet business goals. This era of IT planning has been defined as *strategic alignment*. The strategic alignment model combines the traditional notion of functional integration with the concept of strategic fit (Henderson and Venkatraman, 1989). Today, business and IT strategy are viewed as critical aspects of the firm's overall strategic position (Eisenhardt and Schoonhoven, 1996).

The Strategic Alignment Model

The strategic alignment model (Henderson and Venkatraman, 1989) is composed of four domains: (1) business strategy, (2) organizational infrastructure and processes, (3) IT strategy, and (4) IT infrastructure and processes. Each domain is composed of three components. These 12 components determine the extent and type of alignment within a corporation.

The concept of strategic alignment is based on two building blocks: strategic fit and functional integration (Figure 7.1). The *strategic fit* axis recognizes the need for any strategy to address both external and internal domains (Henderson et al., 1996). The external domain for business strategy is concerned with business scope, distinctive competencies, and business governance. In contrast, the internal domain for business strategy is concerned with the administrative structure, business processes, and the development of human resource skills.

The strategic alignment model proposes that an IT strategy should be defined in terms of an external domain (how the firm is positioned in the IT marketplace) and an internal domain (how the IT infrastructure is configured and managed). In the

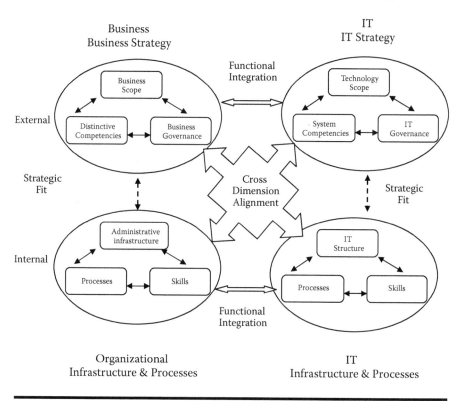

Figure 7.1 Strategic alignment model. (Adapted from Henderson and Venkatraman, 1990.)

external dimension, IT strategy is driven by technology scope, systemic competencies, and IT governance. The internal IT domain is concerned with IT architecture, IT processes, and IT skills.

Functional integration recognizes the dynamic relationship between IT and business strategies. The strategic alignment model identifies two types of integration between business and IT domains. One is at the strategic level, looking at the potential for IT to both respond to and shape business strategy. The other is at the operational level, focusing on the general organizational and IT organizational aspects of the firm.

Given the importance of strategic fit and functional integration, the use of *cross-domain alignment* permits both strategic fit and functional integration to be addressed simultaneously (Papp, 2001). When this is done using a triangle construct, eight cross-domain alignments are possible. Each perspective is unique in terms of the driver or anchor domain and the related conditions. It has been argued (Henderson and Venkatraman, 1990a) that four such perspectives are of particular importance to the discussion of strategic alignment. These are strategy implementation, technology exploitation, technology leverage (competitive potential), and technology implementation.

Strategy implementation is the most common and widely understood perspective corresponding to the classic hierarchical view of strategic management (Gupta and Govindarajan, 1984; Galbraith and Kazanjian, 1986; Bourgeois and Brodwin, 1984; Chaffey et. al., 2000). It reflects the view that business strategy is the driver for both organizational design and IT infrastructure choices (Roth et al., 1991). The technology exploitation perspective is concerned with the exploitation of emerging IT capabilities to impact business scope and influence key attributes of strategy as well as develop new forms of relationships (Armstrong and Sambamurthy, 1999; Bharadwaj, 2000). Technology leverage is a cross-domain perspective that involves developing an IT strategy in response to a business strategy and using the corresponding choices to define the required IS infrastructure and processes (G. Johnson and Scholes, 1993). It seeks to identify the best possible IT competencies through appropriate positioning in the IT marketplace. Technology implementation perspective focuses on the need to build an excellent IS service organization (Henderson and Venkatraman, 1989). This perspective is viewed as necessary to ensure the effective use of IT resources in a growing and rapidly changing world (Bharadwaj, 2000; Powell, 1997).

The strategic alignment model provides a framework to understand the substantive knowledge required to align business and technology strategies. But it does not reflect the dynamic aspect of alignment. Henderson et al. (1996) have identified four management methods for executing, evaluating, and tracking different aspects of strategic alignment. These four alignment mechanisms are (1) value management, (2) governance, (3) technological capacity, and (4) organizational capability. Value management is the organizational mechanism for ensuring that IT resources invested throughout the organization deliver anticipated or greater returns. The governance mechanism specifies the allocation of decision rights for IT activities to the various decision makers within the organization as well as to outside partners. It is not concerned with the day-to-day operational decisions but with the distribution of decision rights, which is consistent with the logic and perspective of strategic alignment. Technological capability deals with the administrative process for creating the required IT capability for supporting and shaping the business strategy. Organizational capability deals with the administrative processes for creating the required human skills and the capability for supporting and shaping the business strategy.

Research that systematically investigates how to strategically align the data warehouses using these mechanisms seems nonexistent. These four alignment mechanisms — value management, governance, technological capacity, and organizational capability — could be useful in aligning the data warehouse to business goals and strategies. Their use for achieving technical integration of the data warehouse is discussed in Chapter 8.

The strategic alignment model also identifies four theoretical attributes of an effective IT planning process as (1) consistency, (2) completeness, (3) validity, and (4) comprehensiveness. The effectiveness of the planning process increases in direct proportion to increases in planning process consistency. Validity is the degree to which a planning process can be systematically biased. A complete and valid planning process will be most effective. Comprehensiveness focuses on the level of detail required to complete the analysis. A consistent planning process can be carried out at various levels of comprehensiveness. The level of detail generated during the process accounts for planning system performance.

The model also reflects the impact of various types of risk on planning effectiveness. Henderson et al. (1996) observe that a major reason for dissatisfaction with the level of integration between business and IT domain, and possibly the absence of value derived from IT investments, is the lack of understanding of the enabling strategic choices that bind a business strategy to the IS infrastructure. This may also be true in the case of data warehouse implementations; a large number of data warehouse implementations have been known to fail (Hwang et al., 2004; Wixom and Watson, 2001).

Developments in Strategic Alignment Research

Various attempts have been made to improve the understanding of the concept of IS strategic alignment and to extend it. Instruments to measure IS strategy and IS strategic alignment have been developed and evaluated (Chan et al., 1997; Chan, 1996). These instruments have been used within organizations to examine existing IS strategy and to highlight IS alignment "gaps." Findings lend support to the view that examining isolated components of strategy and performance can be misleading. Companies with high IS strategic alignment seem to be better performing companies. Alignment between business and IS strategic orientation is linked to IS effectiveness and business performance.

N. Shin's research (2001) considers business strategies in conjunction with IT investments in the analysis of financial performance. His research provides empirical evidence for the importance of aligning IT with business strategies such as vertical disintegration and diversification. The study shows that investment in IT does not in itself ensure profitability for an organization. IT can improve business performance when used in conjunction with vertical disintegration and diversification. Although IT is an essential component (it facilitates better coordination and productivity), it is not sufficient in itself and should be coupled with organizational changes. Increased IT spending improves net profit but not performance ratios such as return on assets (ROA) and return on equity (ROE). Similarly, Davis et al. (2003), in their studies on competitive advantage, find that payoffs from investments in information technology are difficult to recognize and that a sustained competitive advantage from IT-enabled strategies is difficult to distinguish from temporary competitive advantage. N. Shin (2001) finds that by improving scope economies and coordination, IT can shape appropriate business strategies, and at the same time the economic benefits of IT can be leveraged by such business strategies.

Responding to a call for a more inclusive and comprehensive approach to measuring IT business value, Tallon et al. (2000) focused on process-level measures and found that management practices such as strategic alignment and IT investment evaluation contribute to higher perceived levels of business value. They argue that interest in strategic alignment is especially warranted because firms' inability to realize sufficient value from their IT investments is due in part to an absence of strategic alignment. IT evaluation techniques help firms improve strategic alignment, which in turn contributes to higher IT payoffs. The pursuit of strategic alignment is not the sole responsibility of the IS function. Involvement of business executives in IT investment decisions is important and desired because they, as the main clients of the IS function, will be the ones who benefit the most from being able to direct IT resources to better support the business strategy. Kearns and Lederer (2003), on the other hand, argue that although CEOs highly value IT as a strategic tool, CEO participation in IT planning is weak.

Knowledge sharing in the alignment process contributes to the creation of superior organizational strategies (Kearns and Lederer, 2003; Brazelton and Gorry, 2003). Knowledge sharing that stems from collaborative development of business and IT plans ensures the use of tacit and explicit organizational knowledge (Alavi and Leidner, 1999; Kearns and Lederer, 2003; Glazer 1993). Because this knowledge is firm specific, it is capable of rendering a competitive advantage (Porter and Millar, 1985). For firms dependent on information, processes that assimilate and use information in a superior manner have the potential for creating a sustainable advantage. Information intensity is the significance of the information component in value chain activities (Hu and Quan, 2003; Kearns and Lederer, 2003).

Information is valuable and costly, and knowledge is considered the most important resource; thus sustainable competitive advantage lies in what employees know and how they apply that knowledge to business problems (Glazer, 1991).

Glazer (1993) argues that the capacity to recognize information as a firm's primary strategic asset becomes the mechanism through which firms are able to link IT strategy to business strategy and implement a strategic alignment concept. Studies by Kearns and Lederer (2003) provide an explanatory framework of the alignment–performance relationship within the context of a resource-based view and furnish several new constructs. They show that information intensity is an important antecedent to strategic IT alignment, that strategic IT alignment is best explained by multiple constructs that operationalize both process and content measures, and that alignment between the IT plan and the business plan is significantly related to the use of IT for competitive advantage.

Researchers have also addressed the issue of flexibility in strategic planning. The securing and maintaining of alignment between business strategies and IS strategies is frequently cited as a critical concern of IS managers (Burn, 1996). Managers are continually confronted with new and ever-changing competitive pressures from deregulation, globalization, and convergence of industries and technologies. Strategy and strategic planning need to embrace greater flexibility to nurture creativity and innovation (Loebbecke and Wareham, 2003). A rigid information technology infrastructure will frustrate even the best strategic initiatives, making it difficult to introduce change in cost- and time-efficient ways (Prahalad and Krishnan, 2002). Flexibility is particularly important for decision-support applications (Wixom and Watson, 2001). One such decision-support application is the data warehouse. A data warehouse needs to be flexible and responsive to business change (Armstrong, 1997).

The gap between emerging strategic direction and IT's ability to support it can be debilitating. The reasons for infrastructure lags are not purely technical. Organizational issues like IT governance and senior managers' approach to IT investment are equally responsible. A shared understanding and a shared agenda between business managers and IT managers is required to create this flexibility. As data warehouses are built to address business problems, it requires careful planning and alignment between the IT department and business users (Gardner, 1998). Collaboration between the two could enhance the chances of successfully building data warehouses. Cooperation and support from executive management could lay the foundation for such collaboration (Gardner, 1998).

Strategic alignment between business and IT can have a positive business impact if an organization's IT components are a part of a well-integrated organizational system. Pollalis (2003), in his research, develops a strategic co-alignment model that examines three types of integration that impact the planning process and overall performance of information-intensive organizations: (1) technical integration, (2) functional integration, and (3) strategic integration. According to

Calvanese et al. (1999), integration is one of the most important aspects of a data warehouse. Data integration (Widom, 1995; Rahm and Do, 2000), source integration (Calvanese et al., 1998), and architectural integration (Nemati et al., 2002) are important issues affecting performance of data warehouses as well. Organizations with consistent levels of strategic alignment process and output perform better than similar organizations when they have a high degree of technical integration and IT-based functional coordination in place. Communication of the strategic direction within the organization is also necessary. Organizational communication can be improved by integrating the various information resources within an organization.

Other researchers, such as Bai and Lee (2003), have investigated the organizational factors that influence the quality of the IT strategic planning process and the organizational mechanisms for success in strategic planning. Organizational mechanisms, like group interaction, knowledge management, organizational learning, and change management, can be integrated into the IT strategic planning frameworks to enhance the effectiveness of such planning. Group interaction incorporates heterogeneous perspectives and reduces conflicts among stakeholder groups. Knowledge sharing, be it tacit or explicit in nature, is necessary to IS strategic planning. Business knowledge, organization-specific knowledge, IT knowledge, and managerial competencies need to be integrated during IS/IT strategic planning (G.G. Lee and Bai, 2003; Bai and Lee, 2003). Organizational learning leads to increased understanding of IS opportunities and constraints and a shared view of IS utilization. Because a data warehouse is used primarily for organizational decision making (Chaudhuri and Dayal, 1997) and has significant organizational impacts (Wixom and Watson, 2001), these organizational mechanisms can enhance data warehouse planning as well.

In the data warehousing environment, the degree to which the data warehouse technology supports the business issues relevant to the organization can influence the acceptance of the data warehouse technology (Chenoweth et al., 2006). A lack of knowledge of a technology could lead to difficulties in using the technology. The perceived availability of the data warehouse development team as a support group enhances the users' understanding of the purpose of the data warehouse and reduces the difficulty in learning how to use the data warehouse (Chenoweth et al., 2006).

Regardless of the IS/IT strategic planning approach chosen by the organization, it has to be modified to fit the organizational environment, its culture and skills. Andal-Ancion et al. (2003), in their studies to locate specific drivers that determine best strategy for competitive advantage, found that organizational learning was involved when implementing new business methods. Caldow and Kirby (1996) in their studies conclude that effective performance results only when business culture is matched to the goals and strategies of the firm. They identified four specific business cultures — entrepreneurial for invention, hierarchical for mass production, partnership for continuous improvement, and modular for mass customization — and argued that when these business culture forms are aligned to appropriate business strategy, effective performance will result.

Alternatives to the Strategic Alignment Model

As shown by the discussions in the preceding section, alignment of IT strategy and architecture, and business strategy and architecture is a critical success factor for modern organizations. An IT architecture that aligns with the business architecture of an organization reduces costs and provides the opportunity for new products and services. A misalignment between IT architecture and business architecture would mean higher costs and a loss of opportunities to competitors.

According to Gardner (1998), building a data warehouse is a careful alignment between IT and business. Therefore, an architectural choice for building the data warehouse driven by business strategy would align the data warehouse more closely to the business strategy and goals. In the case of a data warehouse, the scope of the business vision can dictate the architectural approach (Murtaza, 1998) chosen for the data warehouse, ranging from a data mart to an enterprise-wide approach. An approach that aligns data warehousing technology with an organization's strategic objectives (Weir et al., 2003) would enhance the long-term success of the data warehouse.

Notwithstanding the importance of architecture alignment, practical guidelines for software architects to achieve alignment are still unavailable. Also, research in the information management area focuses almost exclusively on the strategic level. There are no practical design guidelines for the operational level. Van Eck et al. (2004) point out that the merit of the strategic alignment model is in its recognition of an external orientation of IT strategy. However, it is difficult to apply the model in practice, because Henderson and Venkatraman (1990b) do not provide an *operationalization* of their model. Second, they point out that the strategic alignment model is not a constructive theory of strategic alignment, because it does not provide any guidelines on how to reach a specific goal. Given a particular alignment case study, there are no objective, concrete criteria to determine which of the alignment perspectives play a role in the case.

van Eck et al. (2004) present a framework for architectural alignment called GRAAL (Guidelines Regarding Architecture Alignment) which tries to address this problem. Their framework can be positioned between approaches for software architecture and strategic alignment. They show that in modern organizations, architecture at the application level is designed and managed in a different way than at infrastructure level. IT infrastructure is designed at a time when most of its users are not known. The design of the infrastructure is therefore not motivated by user needs but by the IT strategy of the organization. Alignment at the application level is motivated both by end-user needs and by features of the currently available infrastructure. Consequently, a key alignment problem is the alignment of IT infrastructure services to the application needs of business processes. As yet, there are no comprehensive guidelines that will assist the practicing architect in aligning architectures at all levels. There are no "operationalization" of strategic alignment models.

Maes (2000) proposes an alignment model that downplays the importance Henderson and Venkatraman attribute to technology itself. In their framework, the internal level of the strategic fit dimension is split into two levels: structure and operation. This is motivated by the fact that the structure of the operational processes has to be determined first before they can be executed. These two activities occur and are managed in different ways. In the functional integration dimension, the IT level is similarly split into technology and "information and communication." This reveals a difference with Henderson and Venketraman's emphasis on an organization's competencies with respect to information technology. Maes (1999) argues that information, rather than technology, is the real carrier of value and the source of competitive advantage.

Beeson and Al Mahamid (2003) conclude that although the strategic alignment model introduce a fertile set of ideas into the IS field, it has deficiencies. It does not help managers choose the right perspective (from business execution, IT potential, competitive potential, and service level) for a particular circumstance. Also, it assumes a more static general business environment than that which prevails today, given the existence of the Internet and continued intensification of competition. They also point out the need for theoretical development in the view of alignment as a "process" rather than a set of perspectives. Alignment is seen by them as a process requiring continuous adaptation and coordination of plans and goals within a real and shifting framework of interactions and alliances. In their survey of managers' attitude towards strategic alignment, they find that although IT managers understand business needs, business managers do not understand IT. They suggest not only creating frequent communication channels between IT and business managers to facilitate understanding, but also that IT managers should use business language.

Burn and Szeto (2000) contend that effective alignment of IT and business strategies can be attained by means of strategic information systems planning (SISP). Their study attempts to identify factors that contribute to successful alignment. They propose that by coordinating the objectives and views of IT and business managers, companies can outperform these without such alignment. They find that although IT and business managers have similar perceptions with regard to the drivers and the need for alignment principles between business and IT strategies, there is a divergence of view concerning the different problems in implementing the alignment of business and IT strategies. They conclude that the Henderson and Venkatraman (1990b) model does not identify the issues that are relevant in a practical sense, because they are perceived to be in the theoretical model. He lends support to the argument that the theoretical view is not supported by the practical implementation.

A six-step approach has been designed by Luftman and Brier (1999) to make strategic alignment work in an organization. The process mirrors the traditional strategic planning process. The six-step approach involves (1) setting the organizational goals, (2) establishing a team, (3) understanding business–IT linkages, (4) analyzing and prioritizing gaps, (5) specifying actions for project management, and (6) choosing and evaluating success criteria.

It is generally accepted that strategic alignment is difficult to achieve and sustain over time. Ideally, organizations should always have a high level of strategic IS alignment. When an organization needs to change business or IS strategies, all the aspects of strategic IS alignment should also be modified. But organizations sometimes make decisions that take them out of alignment (paradoxical decision), go too far in certain respects (excessive transformation), or reverse a change and go back to the original position (uncertain turnaround). These trajectories of strategic IS alignment have been studied by Hirschheim and Sabherwal (2001), who sought to examine the factors explaining these problematic trajectories of alignment. Organizational inertia, sequential attention to goals, gaps in knowledge, and split responsibilities help explain paradoxical decisions to some extent. Split responsibilities and underestimation of problems play a role in excessive transformations. Organizational inertia, sequential attention to goals, and underestimation of problems seem to explain uncertain turnarounds. These problems may be addressed by taking steps for aiding strategic IS realignment efforts. Knowledge integration across business and IS domains through technical and social approaches is one such step. Shared knowledge between business and IS executives would help avoid paradoxical decisions caused by business executives' lack of IT knowledge and IS managers' inadequate business knowledge. Process integration, such as through the integration of business and IS planning processes, could reduce the adverse effects of split responsibility.

Different authors and researchers have suggested different models of alignments. Smaczny (2001) questions whether alignment between business and information technology is the appropriate paradigm to manage the IT function in today's organization. According to him, a strategic approach that allows for handling a chaotic environment and offers a rapid response is required. The strategic alignment model, because of its synchronization overhead, will not be flexible and responsive enough to deliver necessary outcomes in a fast-changing business environment. Smaczny suggests that a notion of fusion should be used as the new paradigm for integrating the role of IT in an organization. IT strategy should be developed not separately from, but at the same time as business strategy. All the impacts would be evaluated at the same time and the technology forms part of a fully integrated "organic being."

Enablers of Business–IT Alignment

Several frameworks have been proposed in the literature to study and explore alignment between IT and business strategy. However, there is little empirical evidence or a roadmap to carry out alignment. Researchers in this field have largely confined themselves to theoretical issues and practical generalizations. No study has focused on how organizations actually achieve alignment. The crucial issues related to alignment of the data warehouse to business strategies have not been deeply investigated. There is no appropriate route to achieve coordination and cooperation between business strategy and IT strategy. However, based on the preceding discussion, alignment between business strategy and information technology may be facilitated by paying attention to the factors discussed here.

Culture

Firms that are better aligned have cultures in which strong partnerships between business and IT are cultivated at all levels. This is reinforced when business and IT managers are jointly accountable for prioritizing, allocating resources, and delivering major IT investments. Through these relationships, necessary communications can occur to ensure that both business and technological capabilities are integrated into effective solutions at each level of the business. The degree of mutual understanding and cooperation between business and IT managers can be greatly enhanced by communicating in a common language. IT organizations have for several years tended to use technical terms. By using terms business managers can understand, IT professionals can take a more proactive role in helping to educate business managers about current technologies. A climate of clear communication becomes a necessity for alignment to succeed.

Customer Focus

Several studies discuss strategy as embarking on an era of customer focus enabled by technology developments (Loebbecke and Wareham, 2003; Kodama, 2002). Technology is applied to strengthen a company's customer orientation and increase customization of products. With increasing expectations from customers, companies must develop new ways to provide value-added products and services to customers. Proliferation of information, reduction in geographic boundaries, and acquisition of sophisticated knowledge about products and services by customers themselves require business strategies to produce new business models. Kodama (2002) suggests that these new models should consciously take customer knowledge into account and build a strategic community based on a strategic partnership with a core of highly educated and experienced customers.

Organization

Clearly defined roles, responsibilities, and accountabilities are key to successful collaboration between business and IT managers (Bruce, 1998). Commitment by top management and the involvement of IT in corporate strategy formulation are integral to better alignment. Technical skills have always been the preeminent requirement for IT professionals. But the skills that organizations need in their IT staff have assumed new dimensions. Skills in people management are becoming more critical. A new emphasis on negotiation, leadership, listening, innovation, teamwork, customer service, and consensus-building skills is required to succeed. Cooperation among the stakeholders is important in order to reduce potential conflict that may jeopardize the implementation of strategic IS plans. Key coalitions and bases of power within the organization must also be supportive of the process of strategic planning of information systems and its alignment to business strategies. The organizational learning that accompanies the planning experience should result in improved capabilities to achieve alignment between IS and business strategies, and foster cooperation and partnership among functional managers and user groups. Segars and Grover (1998) conclude that alignment may be manifested through an understanding of organizational objectives by top IS planners, a perceived need to change IS objectives in light of changes in corporate strategy, mutual understanding between business and IS managers, and a heightened view of the IS function within the organization.

Joint Responsibility

Alignment between business strategy and information technology strategy requires a strong interdependent relationship between the business and IT managers. IT and business managers need to be jointly responsible for defining alignment and collaborating continuously through strong partnerships and appropriate allocation of resources. IT managers need to be knowledgeable about how new technologies can be integrated into business and understand the strengths and weaknesses of the technology. They need to delineate to business managers the corporate-wide implications of the technology in question and be privy to corporate strategies. Smaczny (2001) sees the role of the CIO moving from that of a strategic partner responsible for expectation management and technology advice to the role of a business visionary responsible for business innovation and utilization of opportunities created by the technology.

Given the complexity of the alignment process, Edwards (2000) contends that the chief executive officer is in the unique organizational position to set the stage for alignment and its subsequent implementation. The understanding, commitment, and involvement of the CEO in the process can ensure the success of strategic IS alignment to business strategy. Long-term success is dependent on the ability of the CEO to act as a catalyst for strategic alignment and the organization to accept its responsibilities as a community.

Rewards

Performance and behavior are influenced by rewards and incentives, which in turn are based on performance measures. It is important that the performance measures be aligned to corporate objectives.

Conclusion

In conclusion, there is no one universal mode to formulate and implement strategic alignment. It is not an easy task. Alignment is a dynamic and complex process that takes time to achieve and even more effort to sustain. The importance of alignment is well known and documented in the literature, but it is not clear how to achieve and sustain it while building and implementing data warehouses.

Because long-term success of the data warehouse depends on the organization's ability to use the data warehouse to fulfill its strategic milestones (Weir et al., 2003), aligning the data warehouse to business objectives and strategies becomes essential. It involves overcoming difficulties in coordination, communication, priorities, and vision on the part of the business and IS managers. Data warehousing strategy is critical in realizing competitive advantage through the effective use of information (Ma et al., 2000). Although data warehouses can provide many benefits to an organization, the direction of data warehousing can veer off course over time and momentum can lag without continued investment of time from the business side (Watson and Haley, 1998).

Failure to align information technology (and the data warehouse) to business strategy may result in an inability to gain credibility with the business and provide proactive rather than reactive services. Chapter 8 explains how strategic alignment principles can be used to align the data warehouse to business strategies and goals for a successful adoption of the data warehouse.

Chapter 8

Strategic Alignment of Data Warehouses

Companies integrate their data into data warehouses to create competitive advantages, capture new markets, improve data quality, and provide better customer service (Figure 8.1). Adoption of data warehouse technology entails huge capital expenditure and a large development time. Yet data warehouse projects continue to have a high possibility of failure (Wen et al., 1997; Hwang et al., 2004).

Interest in strategic alignment of the data warehouse implementation to the business strategy is warranted because of the inability of organizations to realize sufficient value from their investments in data warehouses (Ballou and Tayi, 1999; Strong et al., 1997; Vatanasombut and Gray, 1999). This chapter provides a set of comprehensive factors that can be used to align data warehouses to business strategy and plans. Alignment of the data warehouse strategy to business strategy contributes to the success of data warehouse implementations.

Conceptual Model

This chapter presents factors that facilitate the alignment between business strategy and data warehouse projects. A conceptual model for the strategic alignment of data warehouses to business strategies and goals is presented hereunder. The model identifies the critical factors that successfully align the data warehouse to the business. Positively addressing these factors will help in a successful implementation of a data warehouse.

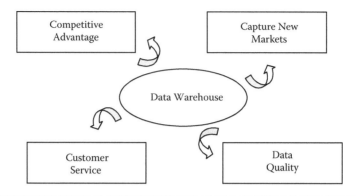

Figure 8.1   Advantages of a data warehouse.

Figure 8.2 illustrates the conceptual model. Factors relating to implementation of technology and to the development process of the data warehouse have been eliminated from this model.

The model shows that data warehouse success depends on strategic alignment of the data warehouse to the business plans and strategies of the organization. Achieving this alignment enables the successful implementation and use of the data warehouse. The factors that influence strategic alignment of the data warehouse to business are (a) joint responsibility between business and data warehouse managers, (b) alignment between business plan and data warehouse plan, (c) business user satisfaction, (d) flexibility in the data warehouse framework, and (e) technical integration of the data warehouse.

These factors in turn are influenced by other attributes, as shown in the model. Senior management's commitment and involvement are expected to play an important role in establishing joint responsibility between data warehouse and business managers. Architectural alignment, knowledge sharing, and good communication between business and data warehouse managers improve the alignment between data warehouse and business plans. Ease of use, data quality, and perceived usefulness enhance business user satisfaction.

Strategic alignment between the business and the data warehouse can be achieved by ensuring that the underlying critical factors are addressed. As shown in Figure 8.2, five factors contribute to successful alignment of the data warehouse to the business plan and strategy:

1. Joint responsibility between data warehouse and business managers
2. Alignment between data warehouse plan and business plan
3. Business user satisfaction
4. Flexibility in data warehouse planning
5. Technical integration of the data warehouse

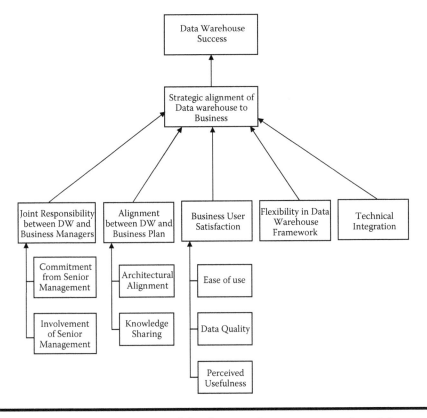

Figure 8.2   Strategic alignment of the data warehouse conceptual model.

The impact of strategic alignment on the successful adoption of data warehouse technology will be examined and analyzed along these five factors.

Joint Responsibility between Data Warehouse and Business Managers

The level of joint responsibility between business and data warehouse managers is critical to strategic alignment and successful adoption of the data warehouse. Commitment by senior management and the involvement of IT in corporate strategy formulation is integral to better alignment. Alignment between business strategy and information technology strategy requires a strong interdependent relationship between the business and IT managers. IT and business managers need to be jointly responsible for defining alignment and must collaborate continuously through strong partnerships and appropriate allocation of resources. IT managers need to be knowledgeable about how new technologies can be integrated into

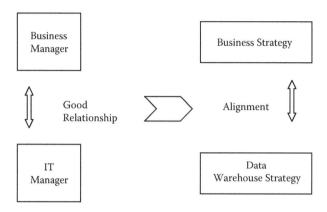

Figure 8.3 Alignment of the data warehouse.

business and must understand the strengths and weaknesses of the technology. (See Figure 8.3.)

Champions and senior management support are important to data warehousing projects, because they actively promote the project and provide information, material resources, and political support. Support and commitment from the senior management is important to secure the required capital, human support, and internal resources during the adoption and development process. (See Figure 8.4.)

Data warehouse initiatives are often IT driven. As a result, commitment, resources, and management buy-in are difficult to obtain. A business-driven warehousing initiative has a better chance to succeed because it has organizational commitment. The data warehouse design and scope is related directly to the business strategy and it is managed as a business investment.

Data warehouse projects require a strong mandate from senior management. Problems arise in successful adoption of data warehouses because of inadequate

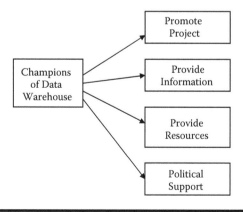

Figure 8.4 Champions of the data warehouse.

commitment and support from senior management. A strong mandate from senior management is needed to impose standards, because different areas within a company may resist changing their ways.

Data warehouse projects are usually large in scale, and funding problems often arise. The benefits derived from a data warehouse are not always foreseeable or quantifiable. Some benefits are intangible and cannot be traditionally cost-justified or measured. The data warehouse is usually viewed as a cost because the data by itself is of limited value. It is the decisions based on the data provided by the warehouse that provide value and benefits. A *strong business case* is therefore required for data warehouse projects.

Strategic alignment of the data warehouse depends as much on organizational commitment as teamwork. Groups of people at all levels engaged in the warehousing efforts, be it the data warehouse stakeholders or the front-line programmer, have to work as a team. Business and IT have to work together as one team toward a common goal. The absence of alignment between the data warehouse and business is one reason many data warehousing efforts fail to realize value from their investments. Acquiring a holistic concept of data warehousing and understanding of the problems by the business would reduce the misalignment between the business and data warehousing initiatives. Increasing business knowledge and knowledge of business policies on the data warehousing team side would further help reduce the misalignment issue. A highly well-aligned data warehouse initiative will still need the support of organizational strategy and a culture to succeed. A close relationship between the organizational strategists, CEO and CIO, is imperative for the success of an enterprise-wide data warehouse. (See Figure 8.5.)

The association of senior management and its involvement with successful implementation of IS has been pointed out by researchers in information systems as well. Tallon et al. (2000) asserted that the pursuit of strategic alignment was not the sole responsibility of the IS function. Involvement of business executives in IT investment decisions was important and desired because they, as the main clients of the IS function, would be the ones who would benefit the most from being able to direct IT resources to better support the business strategy. This is especially

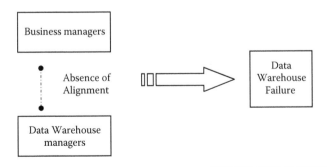

Figure 8.5 Absence of alignment.

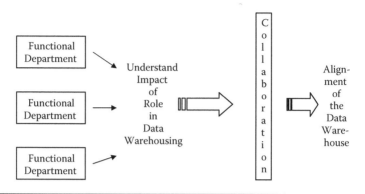

Figure 8.6   Collaboration.

pertinent to data warehouse projects, because they are often built as a strategic or competitive tool for business managers.

An enterprise-wide data warehouse would have an impact at various employee levels and work process levels. Comprehension by individuals of the impact of their role in the organization as well as in the data warehousing process would enable a more effective accountability structure and ownership of the data warehouse. An accountability structure based purely on the organizational structure may not always succeed. Cross-departmental collaboration and alignment is necessary for the data warehouse to succeed. Recruiting the right team members in the data warehousing process who complement the organizational culture is one way of assuring collaboration between teams. (See Figure 8.6.)

Senior leadership style, communication, and culture impact the way data warehouses are executed. In the past, organizational strategy was the responsibility of upper management. With the advances in information technology, data warehouses enable strategic decision making at lower levels within the organization. It makes available enterprise-wide, consistent, and timely information for decision making. When a data warehouse is strategically aligned to the business, it changes the utilization of information technologies toward organizational goals. In turn, benefits from such an investment ensure the success of the data warehouse.

Competent data warehouse leadership from the IT side is required to align the data warehouse to business goals. Building strong relationships between the data warehouse team and the business users is required. This promotes interpersonal communication and alignment with business requirements. Priority recognition leads to realization of the value of the information in the data warehouse by identifying opportunities to utilize and leverage the data warehouse. Strong communication channels reinforce strategic alignment of the data warehouse by raising the level of openness between IT and business leaderships. Creation of a common language between data warehouse and business leadership enhances cooperation toward attaining organizational goals. (See Figure 8.7.)

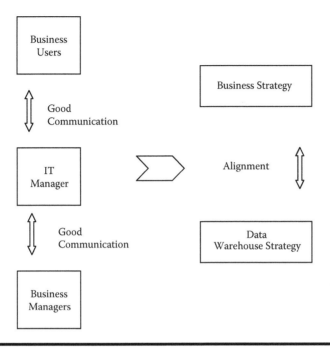

Figure 8.7   Good communication.

In the data warehouse environment, performing joint application development sessions with senior management helps in reaching an understanding of the business plan and opportunities. This facilitates the choice of the appropriate subject areas in a data warehouse and also identifies the ones with the best potential return on investment. Participation by all stakeholders would ensure that everyone's input and perspective are taken into consideration.

Although IT and business managers have similar perceptions with regard to the drivers and need for alignment principles between business and IT strategies, their views diverge concerning the different problems in implementing the alignment of business and IT strategies. In data warehouse projects, although IT managers may generally understand business needs, business managers may not understand IT. Creating frequent communication channels between IT and business managers facilitates this understanding.

For strategic alignment of data warehouses, communication of the strategic direction within the organization is necessary. Given the complexity of the alignment process, the CEO is in the unique organizational position to set the stage for alignment and its subsequent implementation. The understanding, commitment, and involvement of the CEO in the data warehousing process can greatly improve the prospects for strategic alignment of the data warehouse to business strategy. Long-term success of the data warehouse depends on the ability of the CEO to act

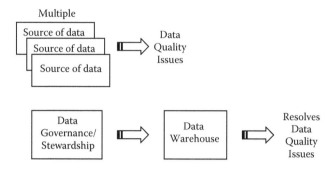

Figure 8.8   Data quality.

as a catalyst for strategic alignment and the ability of the organization to accept its responsibilities as a community.

The issue of *data quality* is pertinent to the success of the data warehouse. Source data is usually dispersed across many different systems across business units in the organization. The complexity of the data becomes evident when data is loaded into the data warehouse for integration. Data issues relating to inconsistencies in naming conventions, data currency, data fragmentation, and duplication are experienced. Identifying the best quality of data to be loaded into the data warehouse requires time and a concerted effort from data warehouse professionals as well as the end users of the system. (See Figure 8.8.)

In a data warehouse, the data is not only integrated across different functional units of the organization, but also includes external entities such as customers and suppliers. As data warehouses evolve, myriad data quality issues emerge. The issue of data quality needs to be viewed from a business-wide sense. The data stewardship program can ensure good quality data. Data stewardship would require the engagement of senior management and the building of cross-functional teams comprising business and technical sides. By increasing the business user's awareness of what data exists and where it can be found, active use of the data would be more likely. Moreover, the technical staff would gain insights that may allow them to fully align their work priorities with the company's overall business strategy.

In an enterprise-wide data warehouse implementation, data is shared across business units. The *data steward* plays an important role in addressing data ownership issues. The ownership of the data model by the business is important. The data steward acts as the business owner of the data and assumes the responsibility for assuring the quality of the data. When a problem with data is discovered, the data steward executes a plan to address the it and informs the people responsible for the quality of the data. The data steward helps to define the data and identify the required data within the business.

Strategic alignment of the data warehouse to business goals enhances the effectiveness of the data warehouse from the user standpoint, thus rendering the data

Figure 8.9   Misalignment.

warehouse successful. Operational systems aim to increase efficiency, and this has been the focus of various information systems. Efficiency is related to using the available resources in a superior manner. Effectiveness focuses on achieving the goals. It measures the degree to which the user requirements are met. Organizations have great expectations from enterprise-wide data warehouses, aiming to improve competitiveness in the marketplace and open up new opportunities. These expectations are usually not met when the focus is placed mostly on improving efficiency by integrating the information systems. Effectiveness can be improved by focusing on the goals and needs of the organization and the organization structure.

To attain effective strategic alignment of the data warehouse to business goals, a real understanding of the purpose, benefits, and constraints of the data warehouse and the business environment has to be reached. Alignment of the organizational resources with the data warehouse strategy is also necessary for execution of the strategy. A major risk of misalignment exists when there is a large gap between the conception of the strategy and the process of implementing it. Achieving strategic alignment of the data warehouse to business goals can leverage business plans and help in obtaining competitive advantage. (See Figure 8.9.)

Usually, in organizations, the business strategy drives the IT strategy. Business strategy leads to business planning and IT planning. Strategic alignment of the planning process is especially important when adopting new technologies such as data warehousing, because these new technologies can drive the business strategy. For example, some strategic goals that a data warehouse can achieve (as seen in companies like Amazon, Orbitz, or Ebay) are

- Establishment of barriers for competitors
- Influence or alteration of the bargaining power of the organization (in relationship to buyers/suppliers)
- Change of the basis of competition (based on cost, differentiation, or focus)
- Generation of new products

It is generally accepted that strategic alignment is difficult to achieve and sustain over time (Boddy et al., 2005). Organizations sometimes make decisions that take them out of alignment. These trajectories of strategic IS alignment have been studied by Hirschheim and Sabherwal (2001). They found that organizational inertia, sequential attention to goals, and underestimation of problems explained some of those trajectories. They suggested that the problems could be addressed by taking steps to aid strategic IS realignment efforts. Knowledge could be integrated across business and IS domains through technical and social approaches. Shared knowledge between business and IS executives would help avoid paradoxical decisions caused by business executives' lack of IT knowledge and IS managers' inadequate business knowledge. Hirschheim and Sabherwal (2001) reported that process integration, through the integration of business and information systems' planning processes, could reduce the adverse effects of split responsibility.

Strategic alignment of the data warehouse depends on a *strong relationship between the data warehouse managers and the business executives*. For the enterprise data warehouse to be successful, a joint partnership is required between the business and IT managers. Data warehouse and business managers should be jointly committed to the success of the data warehouse. There should be shared responsibility for the outcomes. Organizational structures or procedures that foster a good relationship between data warehouse managers and business managers allow for a continuous and productive partnership. Obtaining needed leverage in large organizations is a challenge, but it can be overcome through the involvement of business managers in the data warehousing process. Organizational cultures that foster closeness between managers along the organizational hierarchy foster a continuation of good relationships and cooperation. Interventions that promote collaboration and partnership lead to effective governance of the data warehouse.

Alignment between Business Plan and Data Warehouse Plan

Data warehousing strategy is impacted by the business strategy of the organization and impacts the business strategy in return. The achievement of alignment between the two – data warehousing and business strategy—is by integrating the data warehouse and business plans. Data warehouse plans must support the business plans (Figure 8.10).

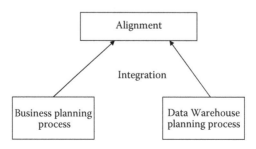

Figure 8.10 Alignment between business plans and data warehouse plans.

Understanding how to align the data warehouse to organizational goals is a huge challenge, but it is capable of producing significant payoffs. Synchronization between the data warehouse plans and the business plans can result in huge benefits. Creating value at the business level using information technology has been a subject of study and interest for a while. Most managerial studies on strategy focus on business-level strategies for competitive advantage. Creating value at the enterprise level by aligning enterprise-wide projects like data warehouses to business strategies has not received as much attention. Alignment of the data warehouse to business strategies involves many factors. They consist of building consensus at the senior management level, mobilizing the senior management at both the business end and the IT end, communicating the strategic business and data warehouse plans to the lower level teams, securing resources for the data warehouse plans, and communicating and integrating the business plans into the data warehouse plans at all levels.

The degree of alignment between business and data warehouse plans is critical to strategic alignment and successful adoption of the data warehouse. Business strategy is the driver for both organizational design and IT infrastructure choices. Developing a data warehouse strategy, in response to a business strategy, and using the corresponding choices to define the required data warehouse infrastructure and processes, brings about closer alignment. It exploits the emerging data warehouse capabilities to impact business scope and influence key attributes of strategy. Building a data warehouse that answers the needs of the business user and provides a high return on investment would be more likely to bring it into alignment with the business.

Aligning the data warehouse plans to the business plans creates synergy and adds value to the business. For synergies to occur, both business managers and data warehouse managers have to play an active role. They have to identify and coordinate opportunities to integrate the data warehouse plans with the business plans. Strategic alignment of data warehouse plans will create greater value for the organization than what can be achieved by planning independently. Effective governance and communication reduces the risk of misalignment. Adopting processes that enable the senior managers of the business and data warehousing group to work together to reach consensus about the business plans and data warehouse plans builds understanding and trust across organizational boundaries. Joint responsibility for the overall alignment provides an opportunity for synergy to be created. Whenever plans are modified at the business level, the data warehouse plans need to be realigned accordingly.

The decision to build a data warehouse and selection of its subject areas must be based on sound business fundamentals. The organizational vision, goals, business plan, and strategic vision could provide a roadmap for the data warehousing team. Selecting a data warehouse project should be a careful tradeoff between business and technology perspective, IT preparedness and business needs, and present technology infrastructure and future infrastructure (Figure 8.11). Data warehouse

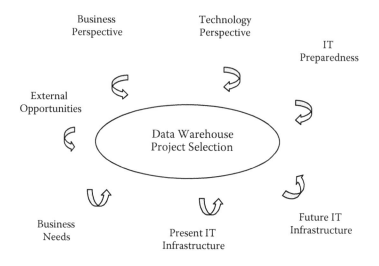

Figure 8.11   Factors affecting data warehouse project selection.

planning and management is a continuous process and should be flexible to adjust to external opportunities for business strategies.

Alignment of IT strategy and architecture, and business strategy and architecture is a critical success factor for modern organizations (Figure 8.12). An IT architecture that aligns with the business architecture of an organization can reduce costs and provide the opportunity for new products and services. Misalignment between IT architecture and business architecture could mean higher costs and a loss of opportunities.

In the data warehouse environment, the scope of the business vision can dictate the architecture approach. A short-term vision requires a lower budget, quick return on investment, and implementation with a small resource requirement, as offered by data marts (Figure 8.13). More strategic objectives of long-term gain and full organizational control necessitate the enterprise data warehouse architecture (Figure 8.14). The most popular architecture formats from which to make choices are the enterprise warehouse model, operational data store, DSS data warehouse, and the data mart.

Poor alignment of data warehouse plans to business plans results in costs, both time and resources, and does not deliver expected results. It is difficult to achieve strategic alignment. Business leadership is not always aware of the complexities involved in planning an enterprise-wide data warehouse due to their limited exposure to information technologies. Many data warehouse initiatives fail due to the complexity in the process and poor planning. Alignment can be achieved by partnering the data warehouse leadership with the business leadership and engaging the business in the planning processes through open communications. Poor alignment

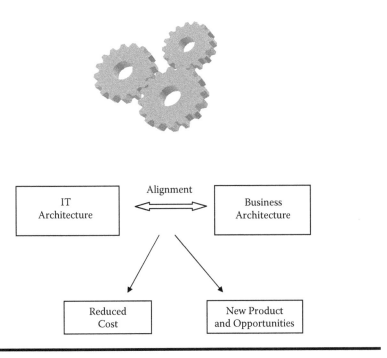

Figure 8.12   Benefit of IT — business architecture alignment.

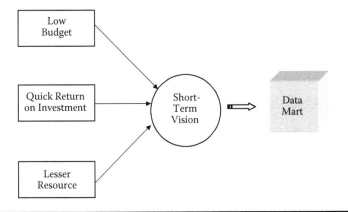

Figure 8.13   Short-term vision.

would decrease as both IT and business increase their understanding of the common objectives of the data warehouse.

An enterprise-wide data warehouse spans data across many different functional units within an organization. Coordination between departments can be time-consuming and inefficient. Enterprise data warehouse implementations require a

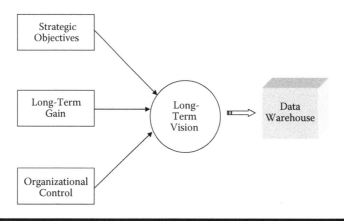

Figure 8.14   Long-term vision.

culture of teamwork and knowledge sharing. Good communication and sharing of information can help motivate action toward meeting the users' needs and expectations. If departmental plans are formed reflecting the overall business strategies and not in isolation, achieving integration of data for the enterprise-wide data warehouse is greatly facilitated. Management processes that are well aligned help reduce difficulties while implementing enterprise-wide solutions. This alignment needs to be managed and sustained on an ongoing basis. For effective data integration, the alignment process requires the cooperation of individuals from across the different functional units. Assigning accountability to individuals or committees and managing the alignment process proactively help address this problem to a large extent. (See Figure 8.15.)

Business User Satisfaction

The degree of business user participation and satisfaction is critical to strategic alignment and successful adoption of the data warehouse. Recent discussions on strategy describe it as embarking upon an era of customer focus enabled by technology developments (Loebbecke and Wareham, 2003; Kodama, 2002). In the data warehouse environment, the business users are the main customers of the system. A basic requirement for a successful data warehouse is its ability to provide business users with accurate, consolidated, and timely information. The strength of the data warehouse is its ability to organize and deliver data in support of management's decision-making process. It can support business decision making by integrating data from multiple, incompatible systems into a consolidated database. The transformation of data into meaningful information allows employees to perform substantive, accurate, and consistent analyses.

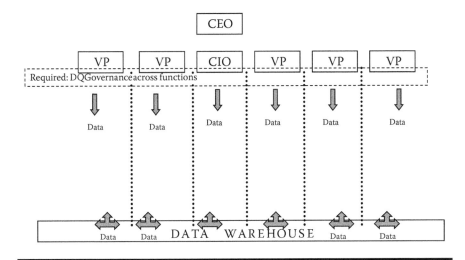

Figure 8.15   Cross-functional dysfunction.

For firms dependent on information, processes that assimilate and use information in a superior manner have the potential for creating a sustainable advantage. Sustainable competitive advantage lies in what employees know and how they apply that knowledge to business problems. The data warehouse is capable of rendering competitive advantage because the data it contains is firm specific.

Business user satisfaction is based on meeting user requirements (Figure 8.16). Defining user requirements for an enterprise-wide data warehouse consists of defining both the organization's data warehouse strategic and operational requirements. A successful data warehouse requirement definition process must encompass both a validation and a prioritization of the user requirements. Because the data warehouse is enterprise-wide, multiple stakeholders across different structural levels need to be recognized. Proactive management of stakeholders' expectations during requirements gathering is important for business user satisfaction.

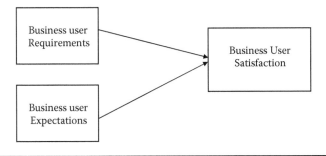

Figure 8.16   Business user satisfaction.

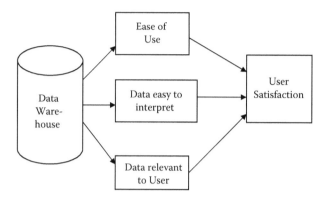

Figure 8.17   User satisfaction.

Understanding the data that is in a data warehouse is a cornerstone to the success of the data warehouse. A primary cause of failure of data warehouse projects tends to be misunderstanding of the data in it. The data that is stored in the data warehouse needs to be easy to interpret by the business user. A data warehouse should not be simply a very large database. In addition to being a database, it needs to provide the information necessary to answer business questions. It needs to do this in a manner that is comfortable and intuitive to the business user. The competitive advantage of a data warehouse, thus, would increasingly depend on the bulk of the organization's employees being able to quickly and easily access the data and interpret the information. (See Figure 8.17.)

End-user participation has a direct impact on the adoption of data warehouse technology. Selection and inclusion of appropriate users in the project team is important. User participation is essential for better communication and coordination of the users' needs. Additionally, end-user participation can help manage users' expectations and satisfy user requirements.

User involvement during user requirements definition can significantly impact user satisfaction. Emotional complexity is often introduced during requirements definition because it is often difficult to manage all stakeholders' expectations and maintain their continued support. Gathering requirements enables the stakeholders in the data warehouse as well as the users to communicate the purpose and set the expectations from the data warehouse. Users and stakeholders often express their needs as overall expectations from the data warehouse and how it could improve their business. The *business requirements* describe the goals and expectations of the stakeholders, which form the basis of the data warehouse requirements. Business requirements represent the top level in the requirements chain. They convey business opportunities and objectives, organizational requirements at a high level, and value to be provided by the data warehouse to the organization. These expectations from the business stakeholders then need to be further refined into detailed and complete specifications for the data warehouse development team.

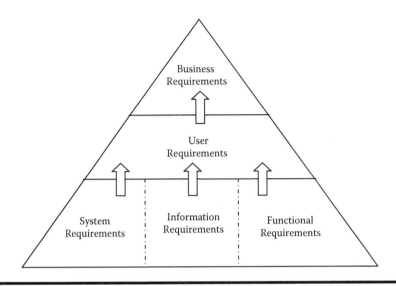

Figure 8.18 Types of requirements.

User requirements gathering is the next level in the requirements chain. It is collected from the actual users of the data warehouse. It describes the tasks the users must be able to perform and accomplish using the data warehouse. It also describes the nonfunctional features of the data warehouse the users require to successfully use the data warehouse and for the data warehouse to be well accepted by the user community. (See Figure 8.18.)

These user requirements form the basis for the detailed system requirements, functional requirements, and information requirements of the data warehouse. The system requirements must align with the user requirements and business requirements to achieve business and user satisfaction.

Success is determined by the user using the data warehouse and achieving high levels of user satisfaction. Involvement of the data warehouse stakeholders and users in the definition of requirements reinforces the intent of the users to use the data warehouse. Involvement of the stakeholders brings about consistency between business objectives and requirements definition of the data warehouse. This also brings about reinforcements among the requirements-gathering activity and produces compatibility between business strategy and data warehouse efforts and objectives. (See Figure 8.19.)

User-related factors such as user participation, user training, and user acceptance are important to the success of a data warehouse system. For a data warehouse to be accepted by its end users, not only must the system be perceived as useful and easy to use, but it is also important that the end users perceive the system to be compatible with their values and past experiences and to be a good fit with the organizational context. User training is a significant factor for user participation,

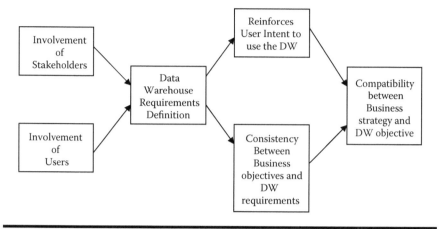

Figure 8.19 Stakeholder and user involvement.

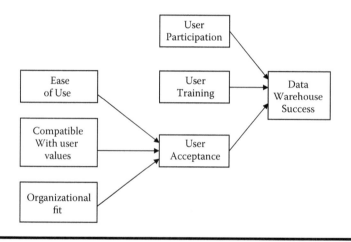

Figure 8.20 User-related factors affecting data warehouse success.

and promotion of user or developer communication during the system development process reduces user conflict. (See Figure 8.20.)

One area of contention and a continual source of friction that ultimately leads
to user dissatisfaction is the recognition of the ownership of the data warehouse.
Ownership is translated into the responsibility for the quality and integrity of the
data as well as their "fitness for purpose." Responsibility has to be clearly delineated
for the group defining the required fields, developing the data model, integrity controls, and production of the data feed. It would also involve business and functional

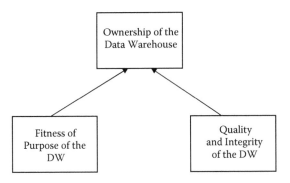

Figure 8.21   Ownership of data warehouse.

requirements documentation and user support. A formal planning process through which business data requirements are captured and escalated in a timely manner would avoid several user issues down the path, yielding satisfied users. (See Figure 8.21.)

Business user satisfaction is also dependent on the user experience of the data warehouse. Involving the business user in all phases of building the data warehouse reinforces the intent of the user to use the data warehouse. The promoters of the data warehouse are to be found at different levels of the organization and should not be restricted only to upper management. Engagement of mid- and lower-level management in discussions and data warehouse planning can facilitate easier user acceptance of the data warehouse.

Failure to align the data warehouse to the business strategies may result in an inability to gain credibility with the business users. Building a strategic community based on a strategic partnership with a core of highly educated and experienced business users should help align the data warehouse to the business. For a data warehouse to be successful, it is important to quickly build trust in the system among the users at the start of the project and continue outreach among the users throughout the implementation and training period.

Technical Integration

Henderson et al. (1996) have identified four management methods for executing, evaluating, and tracking different aspects of strategic alignment. These four alignment mechanisms are (1) value management, (2) governance, (3) technological capacity, and (4) organizational capability (Figure 8.22). The following discussion uses these mechanisms to find ways of achieving technical integration in data warehouse projects. Studies in IT have been used as a starting point for the discussion because data warehouses have a significant IT component.

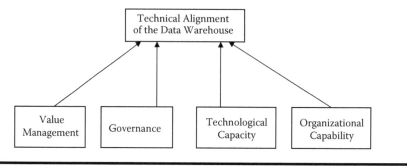

Figure 8.22 Technical alignment of data warehouse.

Value Management: Value management is the organizational mechanism for ensuring that IT resources invested throughout the organization deliver anticipated or greater returns. Previous studies show (Henderson et al., 1996) that investment in IT does not in itself ensure profitability for an organization. Selecting technology for its own sake is a costly and often dangerous decision. Hence, it would seem appropriate that in a data warehouse, technology be selected based on its ability to address business and user requirements. An approach to building a data warehouse that may lead to failure would be to start with tool selection before defining business needs. This approach may fail because interactions between the business and the development team may not have resulted in comprehensive appreciation of business needs.

Responding to a call for a more inclusive and comprehensive approach to measuring IT business value, Tallon et al. (2000) found that management practices such as strategic alignment and IT investment evaluation contributed to higher perceived levels of business value. IT evaluation techniques helped firms improve strategic alignment, which in turn contributed to higher IT payoffs. In a data warehouse project, evaluating technology after the business problem has been identified saves time and resources and allows companies to focus on developing business solutions, and not just technology architectures.

One of the important determinants of new technology acceptance is the perceived ease of use and perceived usefulness. It has been observed (Gorla, 2003) that, despite the potential benefits of data warehousing, corporations often do not provide tools to end users that can be used easily, resulting in non-utilization of tools, millions of dollars of unused software, and unrealized return on investment.

Nah et al. (2004), in their investigation on end users' acceptance of enterprise systems (e.g., data warehouse), found that factors such as perceived compatibility, perceived ease of use, and user attitude were significant determinants in the adoption of a system (Figure 8.23). They found that in order to create positive acceptance among end users, organizational interventions need to focus on the issues of compatibility and technology fit with organizational context.

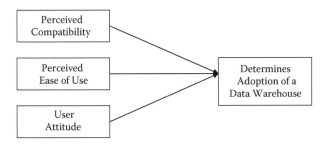

Figure 8.23 Data warehouse adoption determinants.

If a selected data warehouse project requires too much new technology, it may fail due to the inability of the IT department to deploy a large amount of new technology. New technologies often bring new paradigms for understanding how they could fit into the business world. It appears that failure to address all aspects of the technological assimilation process is part of the reason many data warehouse projects fail, resulting in poor value management.

If a data warehouse project with too little new technology is selected, it is possible that the data warehouse may be seen as being unresponsive to user needs. On the other hand, if a data warehouse project is selected that will deliver large amounts of new information to the business, it may collapse under its own weight. Also, a data warehouse project that delivers too little new information to the business can again be seen as being unresponsive to user needs. In either case, it provides poor or no value.

Governance Mechanism: The governance mechanism specifies the allocation of decision rights for IT activities to the various decision makers within the organization, as well as to outside partners. It is not concerned with the day-to-day operational decisions but with the distribution of decision rights that is consistent with the logic and perspective of strategic alignment. In a data warehouse project, the pursuit of strategic alignment should not be the sole responsibility of the IS function. Involvement of business executives in data warehouse investment decisions would be important and desired. As the main clients of the data warehouse, business executives will be the greatest beneficiaries because they are able to direct resources to better support their business strategy.

Technological Capability: Technological capability deals with the administrative process for creating the required IT capability for supporting and shaping the business strategy. Pollalis (2003) in his research found that strategic alignment between business and IT can have a positive business impact if an organization's IT component is part of a well-integrated organizational system. This is also the case with a data warehouse.

Organizational Capability: Organizational capability deals with the administrative processes for creating the required human skills and the capability for supporting and shaping the business strategy. The skills of the data warehousing development team have a major influence on the outcome of the project. A highly skilled project team is better equipped to manage and solve technical problems. Coordination of organizational resources including money, people, and time ensure completion of the project and ultimately affects the adoption of data warehouse technology.

Thus, this discussion shows that degree of technical integration is critical to strategic alignment and successful adoption of the data warehouse.

Flexibility

Many researchers (Loebbecke and Wareham, 2003; Prahalad and Krishnan, 2002) have addressed the issue of flexibility in strategic alignment. They describe alignment as a process requiring continuous adaptation and coordination of plans and goals within a real and shifting framework of interactions and alliances. Managers are continually confronted with new and ever-changing competitive pressures from deregulation, globalization, and the convergence of industries and technologies. Strategy and strategic alignment therefore need to embrace greater flexibility to nurture creativity and innovation. A rigid information technology infrastructure could frustrate even the best strategic initiatives in a data warehouse, making it difficult to introduce change in cost- and time-efficient ways (Prahalad and Krishnan, 2002).

The gap between emerging strategic direction and information technology's ability to support it could be debilitating. The reasons for infrastructure lags may not be purely technical. Organizational issues like IT governance and senior managers' approach to IT investment are equally responsible. Thus a shared understanding and a shared agenda between business managers and IT managers is required to create flexibility.

The issue of flexibility in planning is pertinent to the data warehouse environment. In data warehouse projects, the business strategy could change during data warehouse development. A change in business plan could lead to a shift in the business requirements of the data warehouse. Quick, iterative developments, as opposed to lengthy development projects, address this problem to some extent.

A data warehouse data model and architecture needs to be flexible to reflect the changes in the business process over time. The design of the data warehouse should allow for incremental changes to be managed. The data warehouse technology must have a level of compatibility with the existing technology in the organization. The data warehouse hardware and software should have a level of interoperability with the configuration of the organization's hardware and software, because the warehouse is influenced by the configuration of the existing source systems. The hardware and software infrastructure should try to remain free of conflicts. Because no single technology can satisfy the requirements of a data warehouse, it is advisable to form strategic alliances with a number of vendors.

As business needs change over time, a data warehouse needs to be flexible enough to be responsive to them. If the data warehouse is not flexible enough to adapt to changes in the environment and provide the information users need to run the business, then the organization loses the advantage that the information provides. A data warehouse needs to be built on a foundation that is flexible and responsive to business change. According to Armstrong (1997), this concerns three main areas: (1) the database, (2) the application middleware, and (3) tool integration. In order for the data warehouse to have long-term success, all three areas must have scalability, high availability, and robust manageability. Managing theses changes would span all the components of a data warehouse and would play a vital role in the development and overall success of a data warehouse. To successfully manage a data warehouse, Sen (2004) argues that two conflicting goals need to be managed: maximizing the use of the data warehouse asset while consistently meeting user expectations by continuously monitoring the effects of business change.

Chapter 9

Case Study: Strategic Alignment at Nielsen Media Research

Introduction

Data warehousing technology is relatively new. Most of the data warehouse studies have been done from a technical perspective. The interest in this book has shifted to strategic issues from technical issues. Few previous studies have examined data warehouses from a strategic perspective. This chapter presents the findings from a case study conducted at Nielsen Media Research from a strategic alignment perspective. This case study provides the organizational context for the study of the relationship between data warehouse technology and business strategy and allows for valuable insights.

The first section that follows presents the overview and background of the organization. The second section describes the data warehouse implementation at the organization. The third section describes how the organization achieved strategic alignment between the business objectives and the data warehouse.

Overview of the Company

Nielsen Media Research is the leading provider of television audience measurement and advertising information services worldwide. It is headquartered in New York, with offices throughout the United States. Nielsen Media Research International has offices around the world and provides television and radio audience measurement, print readership, and customized media research services measurement data in more than 40 countries. The company is owned by VNU, a publicly traded international publishing and information leader based in Haarlem, The Netherlands. Nielsen Media Research is part of VNU's Media Measurement & Information Group (MMI), a global leader in information services for the media and entertainment industries. Nielsen Media Research employs around 5,000 people and has a gross turnover of US $1 billion.

The ratings from Nielsen Media Research provide an objective estimate of television ratings and media research. They act as a barometer of peoples' viewing habits. In the United States, Nielsen's National People Meter service provides audience estimates for all national program sources, including broadcast networks, cable networks, Spanish-language networks, and national syndicates. Local ratings estimates are produced for television stations, regional cable networks, cable interconnects, and Spanish-language stations in each of the 210 television markets in the United States.

In addition to providing television audience estimates, Nielsen Media Research provides competitive advertising intelligence services, as well as information on interactive television and Internet usage. Nielsen Monitor-Plus, a service of Nielsen Media Research, is the leader in innovative advertising information services in the United States, providing advertising activity for 18 media including television tracking. Nielsen Outdoor harnesses Global Positioning System (GPS) technology to measure consumer exposure to billboards and other forms of outdoor advertising.

Business Objective

Nielsen Media Research provides high-quality estimates of audiences through a fair and open system that objectively reports viewership. Their customers use Nielsen Media Research's television audience estimates to buy and sell television time. This information is the currency in all the transactions between buyers and sellers, which adds up to approximately $45 million in national and local advertising spending each year.

Nielsen Media Research provides an independent, completely neutral, third-party measurement system, embracing the highest standards of accuracy and integrity in the television marketplace. Assuring the quality of the ratings and the value of this currency is the number one priority for Nielsen Media Research.

Background of the Data Warehouse

The task of audience measurement is a complex one. To capture viewership ratings, Nielsen Media Research has divided the United States into 210 clusters. In different markets, different measurement devices are used. In small markets, diary measurement is used to collect viewing information from sample homes. Paper diaries are mailed out to the households and each household member in a diary sample is asked to write down what channels and programs they watch.

In the larger markets, People Meter, an electronic metering system, is placed in randomly selected households to measure what channel or station is being tuned into and who is watching it. The People Meter is used to produce household and individual audience estimates for broadcast and cable networks and nationally distributed barter-syndicated programs. In medium-sized-market homes, a similar electronic metering system is used to provide set-tuning information on a daily basis and it is supplemented with diaries.

Household tuning and persons viewing data from both the national and local samples for each day are stored in the in-home metering system until they are downloaded to Nielsen Media's computers each night. This information is processed each night for release to the television industry the very next day. Nielsen Media Research collects information from more than 25,000 metered households starting about 3:00 A.M. and processes approximately 10 million viewing minutes for delivery to customers every day.

This data is critical to broadcasters and advertisers alike, who rely on it to make and fine-tune programming decisions and advertising placements. As demand grew for more timely and detailed information, Nielsen Media Research decided to build an online data warehouse that would allow each client to access its vast data store on an ad hoc basis and query that data rapidly. The data warehouse enables the clients to sort through enormous volumes of data to get the intelligence they need to make effective and profitable business decisions.

The Data Warehouse Development

Project Initiation

At Nielsen Media Research, unlike traditional data warehouses that have a support function, the data warehouse is the revenue-earning arm of the business. To address business needs, the data warehouse project was initiated by the CIO and senior management. The data warehouse is tightly aligned to business and was developed by a select team called a Platform team. The team was carefully selected and comprised 25 members. This team brought together people from information technology and product management groups representing the business. The team used an iterative XP methodology to develop the data warehouse, and the data warehouse was available to products as it was being developed. The development of the first data warehouse began in 1997, and two more were built subsequently. The third data warehouse is the subject of this case study.

Data Warehouse Architecture Selection

The data warehouse architecture was selected after considering high-level strategies and detailed-level sources of data. It is a reconciliation of a top-down as well as a bottom-up approach to architecture. The broad vision of the data warehouse had the concepts of data marts in it. It emphasized building the base repository first and then servicing the products through the data marts.

The data warehouse was built after careful evaluation consisting of both technical criteria and business metrics. The data warehouse architecture was selected after an in-house architect researched and explored different options. Different data warehouse database management systems (DBMS) were considered and vendor evaluations were completed by November of 2002. Each of the vendors were quizzed on technical and business metrics and weighted values were given to each criterion. Input was also received from Gartner Group and proof of concept was completed for the products evaluated. Sybase IQ was finally selected over NCR Teradata and IBM DB2.

There were several reasons for selecting Sybase IQ. Because Sybase databases already existed in house for online transaction processing (OLTP)-type systems, the learning period was shorter. The CIO had a good relationship with Sybase and it had his support because it was an existing vendor. Sybase IQ also offered a large, analytical database that could provide fast query response as well as efficient and cost-effective solutions. The data warehouse is accessed using customized software applications as well as the business intelligence tools MicroStrategy and ProClarity.

Physical Implementation

The current data warehouse, which is the subject of this study, is the third-generation data warehouse at Nielsen Media Research. The first was built in 1997 and ran a single application, off a 2-terabyte database. Later, around 2002, a second, larger data warehouse of 20 terabytes was built, supporting multiple applications. A much larger third-generation data warehouse was then built using Sybase IQ. The design of this data warehouse was not completely driven by the applications, as in the earlier data warehouse. (See Figure 9.1.) The data warehouse's architecture allows Nielsen to add new products and applications that can access the audience data warehouse in order to satisfy ever-evolving client demands. This eliminated the need to create duplicate databases for each new application. The data warehouse is currently accessed by more than 2,000 users and is available 24 hours, 7 days a week.

The data warehouse architecture creates base-level structure for the data warehouse by integrating data from all sources that have been identified, including legacy files, mainframe data, ancillary data, and reference hubs through an ETL (extract, transform, and load) process using Synopsis, into the data warehouse. The data is then summarized and is rolled up to separate data marts like Stellar data mart and Overnight data mart, to support different products and applications.

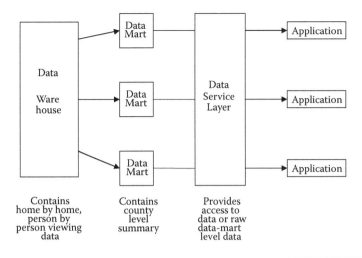

Figure 9.1   Nielsen Media Research third data warehouse architecture.

A separate data service layer insulates the data from the products. It manages the delivery of the data to the applications through Web service, FTP, or subscription service. It also optimizes the data to the product, for example, 24/7 availability or very fast performance. The current data warehouse derives greater value and long-term leverage by repurposing the data and by using better technology and programming architecture. The flexibility in the architecture of the data warehouse allows new products, applications, or data structures to be added to the data warehouse without affecting its performance.

Business intelligence applications run off this data warehouse, enabling syndicators, advertising agencies, and national broadcast and cable networks to understand television viewing habits, resulting in more targeted advertisement placement and optimized revenues. Business intelligence applications use MicroStrategy and ProClarity to access the data warehouse, which contains virtually the entire breadth of Nielsen's data. Clients and research analysts from the TV industry, broadcasting industry, and advertising agencies also use customized and proprietary analytical software to run various analyses on the data warehouse.

Strategic Alignment of the Data Warehouse at Nielsen Media Research

This study shows the impact of strategic alignment of the data warehouse to business strategies and plans on successful adoption of the data warehouse. Strategic alignment between business and the data warehouse can be achieved by ensuring that the underlying critical factors are addressed. The model presented in

Chapter 8 identifies (a) *joint responsibility between business and data warehouse managers,* (b) *alignment between business plan and data warehouse plan,* (c) *flexibility in the data warehouse framework,* (d) *business user satisfaction,* and (e) *technical integration* of the data warehouse as the critical factors influencing alignment and the successful adoption of a data warehouse.

Joint Responsibility between Business and Data Warehouse Managers

Commitment and Involvement of Senior Management

At Nielsen Media Research, the CIO, the senior vice presidents of business divisions, and the product manager were very involved in data warehouse investment decisions. The CIO responded to requests from senior management and initiated the data warehouse project.

All senior business managers at Nielsen Media Research were very involved in the data warehouse investment decision. A senior vice president headed the Platform team, which was responsible for the development of the data warehouse. He ensured the inclusion of the best personnel and resources in business and technical knowledge in the development process. The CIO approved all major requests for investments. The CFO reviewed all investment requests and approved all major requests for investments.

The senior vice president for products was a key decision maker in the plans and approval of budgeting, to make sure the data warehouse had business plans and financial support. The product manager was also highly involved with the data warehouse decisions. If the data warehouse was unable to meet business needs due to any resource constraints, the product manager lobbied for finance and personnel from the business.

The senior management also remained involved in data warehouse decisions at the top level. The CIO reinforced the general technical policy and saw to it that it fit the product strategy and business values. The data warehouse design and data decisions involved cross-functional teams. Because the data warehouse team was given the business plans, all adjustment decisions were made at the data warehouse manager's level.

The association of senior management and its involvement with the successful implementation of the data warehouse helped align the data warehouse to business strategies and needs. The support and commitment by senior management resulted in active promotion of the project and ensured availability of capital, human support, and internal resources during the development and implementation process.

The involvement by business managers and the product manager in the data warehouse is desired and important because, as the users of the data warehouse, they are able to better support their business strategies. This was especially pertinent at Nielsen Media Research where, unlike traditional data warehouses which have a support function, the data warehouse is the revenue-earning arm of the business. Joint meetings between the data warehouse and product manager at Nielsen Media Research helped in reaching an understanding of the business plans and opportunities. It facilitated the identification of areas for development in the data warehouse with the best returns on investment. The data warehouse manager participated in strategy meetings with the business. It provided an opportunity to inform others of the data warehouse architecture and ways to leverage it for strategic purposes.

Involvement of Data Warehouse Managers in Corporate Strategy

At Nielsen Media Research, all levels of data warehouse management were made highly aware of the business plans and corporate strategies by the product manager. The data warehouse managers were aware of the changes in the media industry over the past 5 years and were aware of the technology driving this change in the traditional broadcasting mode. The broadcasters were reinventing themselves to adjust to these changes in technology (for example, iPods, digital video recording [DVR], and video on demand [VOD]) and so was Nielsen Media Research, to stay relevant in this changing industry landscape.

Nielsen Media Research employs various measurement devices in its different measurement markets to capture viewership ratings, from People Meters to paper diaries. New challenges were faced in collecting data from new devices and gadgets such as iPods and cross-media gadgets. Because advertisers are interested in the total picture of what consumers are viewing, Nielsen Media is working toward a single meter strategy in the future. The data warehouse managers were well aware of the biggest challenge for Nielsen Media Research — the ability to service these needs for more flexible and timely reporting as well as providing a wider breadth of data. The data warehouse managers were also aware of business plans to support viewing collection methods for DVR and VOD.

The data warehouse managers were also involved in corporate strategy decisions. The director of IT operations had been the IT representative on the local product strategy team. He had interacted closely with senior management and was also closely tied to the data warehouse project and to the business side. Although the data warehouse manager participated in corporate strategy meetings as needed, the product manager worked with the business managers on all data warehouse issues. She informed the data warehouse manager and project leader of higher level decisions in the company, the business plans and opportunities, and the needs of the business from the data warehouse. The understanding, commitment, and involvement of the product manager in the data warehousing process ensured strategic alignment of the data warehouse to business strategy.

Alignment between business strategy and data warehouse strategy requires a strong interdependent relationship between the business and data warehouse managers. As seen in the previous paragraphs, the product manager, data warehouse managers, and business managers were jointly responsible for aligning the data warehouse to the business by collaborating continuously through strong partnerships and appropriate allocation of resources.

Communication of the strategic direction within the organization is necessary for better alignment. At Nielsen Media Research, good communication between the business and data warehouse managers enhanced alignment of the data warehouse to business strategies. Involvement of cross-functional teams enhanced joint responsibility and integration in the data warehouse design and data decisions. The involvement of business managers and the product manager in the data warehouse was especially pertinent at Nielsen Media Research where, unlike traditional data warehouses which have a support function, the data warehouse was the revenue-earning arm of the business. Joint meetings between data warehouse managers and the product manager at Nielsen Media Research led to better communication and understanding of the business plans and opportunities. Effective communication facilitated the identification of areas for development in the data warehouse with the best return on investment.

Existing business and data warehouse managers can apply these findings to their own situation. In so doing, they can develop an awareness of possible alignment and communication issues and their implications. This awareness will help them to actively manage alignment challenges in the data warehouse and implement corrective steps that enable the strategic alignment of the data warehouse.

Alignment between Business and Data Warehouse Plans

A number of factors contribute to, and facilitate, alignment between business and data warehouse plans. The classic hierarchical view of strategic management suggests that business strategy is the driver for both organizational design and IT infrastructure choices (Henderson and Venkatraman, 1990). Research suggests that developing a data warehouse strategy in response to a business strategy, and using the corresponding choices to define the required data warehouse infrastructure and processes, should bring about closer alignment (Murtaza, 1998). Hence, to gain a comprehensive understanding of the degree of strategic alignment, this section considers (a) architectural alignment of the data warehouse, (b) knowledge sharing, (c) integration of business and data warehouse planning, and (d) communication between business and data warehouse managers.

Architectural Alignment of the Data Warehouse

Alignment of the data warehouse strategy and architecture to business strategy and architecture ensures its successful adoption. A data warehouse architecture that aligns with the business architecture of an organization reduces costs and provides opportunities for new services.

At Nielsen Media Research, alignment of the data warehouse strategy and architecture to business strategy and architecture helped ensure its successful adoption. At Nielsen Media Research, the decision to build a data warehouse and the selection of its subject areas were closely coupled with business needs and business strategies. The scope of the business vision dictated the architectural approach. The organizational strategic plans and objectives provided a roadmap for the data warehousing effort. The data warehouse was built in response to business plans and strategies, to service their needs for greater flexibility and timely reporting, as well as for providing a wider breadth of data.

The data warehouse architecture at Nielsen Media Research was selected after considering high-level strategies and detailed-level sources of data. It is a reconciliation of a top-down as well as a bottom-up approach to architecture. The data warehouse architecture was a combination of applying existing technology investments with tool evaluations to fill in the architectural and technical components. The broad vision of the data warehouse embraced the concepts of first building a base repository and then servicing products through data marts rolled up from the base repository. The data warehouse allows each client to access its vast data store on an ad hoc basis and to query data rapidly. The data warehouse enables the clients to sort through enormous volumes of data to get the intelligence they need to make effective and profitable business decisions.

There have been changes in the data warehouse plans over the past 5 years. Because the content of the data warehouse is closely coupled with the business direction, the changes have been driven by business needs. The data warehouse has adapted to new technology in the industry more quickly than older systems. It has incorporated data from the new measurement device for viewing data, called People Meter. It has also incorporated overnight data into an overnight data mart. In the coming year, commercial data will be added to the data warehouse. Changes have also been made to the integration, collection, and presentation layers of the data warehouse. An iterative development method as well as adoption of Agile methodology has allowed major releases of the data warehouse every 6 months.

Knowledge Sharing

Shared knowledge between business and data warehouse managers helps in avoiding paradoxical decisions caused by business executives' lack of IT knowledge and data warehouse managers' inadequate business knowledge (Hirschheim and Sabherwal, 2001). Knowledge sharing allows one to avoid making decisions that are out of alignment and can help in integrating the business and data warehouse planning process.

At Nielsen Media Research, the data warehouse managers had a good dialogue with the business managers. The data warehouse managers helped the business managers understand the advantages and limitations of the data warehouse. The product manager worked along with the data warehouse and managed the strategic value of the data warehouse. The product manager represented the data warehouse to the business and educated the business on data warehouse possibilities and capabilities.

Additionally, communication between all stakeholders of the data warehouse is essential for its alignment to business strategy and adoption by the business users (Pollalis, 2003; Hwang et al., 2004). Along with knowledge sharing and communication of the strategic direction to the data warehouse managers by the senior management, creating frequent communication channels between data warehouse managers and users is necessary to facilitate better understanding of the data warehouse and its ultimate successful adoption.

Cross-functional teams were highly active in the data warehousing project at Nielsen Media Research, facilitating better understanding of the data warehouse. The data warehouse managers were well informed of business plans and strategies and the data warehouse plans sought to exploit data warehouse capabilities to meet business needs and strategies. Knowledge sharing and the strong and effective communication between the business managers and user on one hand, and the data warehouse manager on the other, brought about closer alignment between the business and data warehouse plans at Nielsen Media Research.

Integration of Business and Data Warehouse Planning

At Nielsen Media Research the data warehouse planning process was well aligned with the business visions and plans. The data warehouse was built in response to business needs. The data warehouse strategy was a response to business strategy and used corresponding choices to define the data warehouse infrastructure and capabilities. Their business and data warehouse planning process were well integrated.

A climate of clear communication becomes a necessity for alignment to succeed. The communication of the strategic direction between business and data warehouse managers facilitated better understanding of the data warehouse and its subsequent successful adoption at Nielsen Media Research. There was significant knowledge sharing among the stakeholders and frequent communication between the data warehouse managers and users at Nielsen Media Research.

Cross-functional teams were highly active in the data warehousing project. The cross-functional team comprised 25 people, consisting of product managers, quality assurance personnel, application and data warehouse developers, data architects, and business analysts. Moreover, a business analyst and a data warehouse person had been assigned for every product being developed, to ensure that the data service could be reused. The product manager for the data warehouse worked with other product managers and senior management to determine their data needs and to balance and prioritize projects.

The product management group at Nielsen Media Research ensured that the business vision drove the design of the data warehouse. Integration of business and data warehouse planning was achieved through data warehouse managers working closely with the product manager, who was involved on both sides. The product manager represented both the IT and the business sides. The role of the product manager is very effective in bringing about better coordination between the business and the data warehouse. The data warehouse managers were well informed of business plans and strategies, and the data warehouse plans sought to exploit data warehouse capabilities to meet business needs and strategies.

The integration of business and data warehouse planning processes was also achieved through portfolio reviews with senior management four times a year. These portfolio reviews drove the IT projects and defined the projects that would be initiated. Product managers looked at high-level strategies and translated them into IT projects and products. The integration of business and data warehouse planning was thus a structured method, stemming from the portfolio reviews. It involves business strategy, prioritization of projects, and planning of resources.

Integration of business and data warehouse planning was also achieved through a process of completing "contract books." The contract book contains the details of the plan, risks, justification, finance, etc., for each major project, when it is undertaken. A contract book was signed for the data warehouse. In effect, it is a reflection that all stakeholders understand the goals, the deliverables, what the project involves, how long it will take, and the expected outcomes.

Integration between business and data warehouse planning processes, knowledge sharing, effective communication between the business managers and users on one hand, and the data warehouse manager on the other, brought about closer alignment between the business and data warehouse plans.

Communication between the Business and Data Warehouse Managers

Communication among all stakeholders of the data warehouse is essential for its alignment to business strategy and adoption by the business users. Along with knowledge sharing and communication of the strategic direction to the data warehouse managers by the senior management, creating frequent communication channels between data warehouse managers and users is necessary to facilitate better understanding of the data warehouse and its ultimate successful adoption.

At Nielsen Media Research, both formal and informal communication channels existed within the organization. Formally, status meetings were held every 2 weeks to review current work in progress. The data warehouse manager and product manager attended these meetings. The strategy team held a meeting once every month, which the data warehouse manager attended. Plans were published to the business and made available to senior management. The CIO received briefings on planning and also occasional status updates.

IT Governance, a tool IT managers and project managers could use to enter all projects, IT staffing, schedules, and priorities, also served as a medium of communication. The product reviews held three to four times a year served as a communication channel as well. At these reviews, priorities were communicated and decisions were made. Frequent, informal communications also occurred between the data warehouse manager and product and business managers, and they met whenever it was needed.

Informally, users communicated directly with the data warehouse project leader through conversations over the phone or in person. The XP methodology adopted for the development of the data warehouse facilitated communication across business and data warehouse units. Formally, if a project needed information from the data warehouse, the project leader would initiate the communication and meet with the data warehouse project leader.

Communications to the data warehouse team were also made through the product manager for the data warehouse. If a project were to impact the data warehouse, if a deliverable was required from the data warehouse, or if any modifications were to be made, it was conveyed to the data warehouse team by the product manager. The weekly meetings held for every project and the daily stand-up meetings also served as channels for communication between the users and the data warehouse team.

At Nielsen Media Research, good two-way communication between the data warehouse managers and its users directly, and also with the facilitation of the product manager along with reciprocal feedback, brought about closer alignment between business and data warehouse plans.

User Participation and Satisfaction, and Data Warehouse Success

A number of factors contribute to the degree of participation and satisfaction of the data warehouse user. This section considers the following four factors: (1) user participation, (2) perceived usefulness, (3) ease of use, and (4) data quality.

User Participation

At Nielsen Media Research, high levels of user participation and satisfaction were found with the development of the data warehouse. User participation resulted in better communication and coordination of user needs and was helpful in managing user expectations and satisfying user requirements.

At Nielsen Media Research users were highly involved in the development of the data warehouse. Some users were involved with the data warehouse from its initial implementation, helping identify low-level details and variables of data to be captured into the data warehouse during the requirements-gathering stage. For instance, users were involved in defining specifications for the data warehouse pertaining to overnight data, ratings data, and program names data. Users were involved in handling instances that did not occur on a daily basis, such as daylight savings time changes, launch of a new network, or changes in network affiliation.

Users were involved with the data warehouse project in defining user needs and validating the output from the data warehouse. Users also loaded local reference data into the data warehouse, and they looked for viewing patterns on the minute-level views of data in the data wareouse using the business intelligence tool MicroStrategy. Users continued to interact with the data warehouse manager to ensure that the required data pieces were included in the data warehouse as it matured. User input also ensured that the data warehouse was aligned to the 3- to 5-year strategies in the business plan.

End-user participation had a direct impact on the adoption of the data warehouse. The selection and inclusion of users in the project team was essential to user satisfaction.

Perceived Usefulness

Users of the data warehouse at Nielsen Media Research found the data warehouse to be supportive of user needs and perceived it as useful.

The data warehouse supported the products and incorporated improvements asked for by the users as well as their clients. It also offered a tremendous amount of flexibility and leverage with future product development as well. Because all the data resided in the data warehouse repositories, the architecture allowed new applications to be added easily. The data warehouse helped to get the products into the market faster and the clients experienced performance gains. Enhancements have been included in the data warehouse, and the user interface for overnight data has become more events driven. This has reduced the clients' support interface with Nielsen Media Research.

The users were satisfied with the relevance of the data warehouse to their day-to-day decisions. The data warehouse adequately supported the users' needs. The data warehouse was accepted and successfully adopted by the users. The response of senior managers to the usefulness of the data warehouse was similar to the business users' perceptions. They found the data warehouse useful in their ability to respond to changes in business direction. The customers of Nielsen Media Research have been provided with information from the data warehouse and have used it to make better business decisions. The Toolbox application available internally within the company has helped in answering to the client base and in giving guidance to the clients. It has been used for customized analysis as well.

The product manager finds that the data warehouse does provide information for the products it supports. The data warehouse supports business by integrating data from multiple, incompatible systems into a consolidated database. Because many legacy systems are still being used within the organization, work is under way to develop an overarching product strategy to rework the customer analysis system so that at a future date it would use the data warehouse exclusively.

Previous research (Triantafillakis et al., 2004; Meyer and Cannon, 1998) suggests that a basic requirement for a successful data warehouse is its ability to provide business users with accurate, consolidated, and timely information. At Nielsen Media Research both senior management and users perceived the data warehouse as useful.

Ease of Use

Generally the users found the data warehouse easy to use. It was easy and transparent for the end user. The user understood the table structure and the data within the data warehouse. Even though the data in the data warehouse was complicated, internal users found it easy to use because of support from the data warehouse team. Some applications running on the data warehouse were intuitive. With the data warehouse, there is better communication and cross-fertilization of ideas across groups. Efforts are under way to improve and have a common look and feel across products.

Users felt that understanding the data warehouse functions and features was getting easier because of the many changes being incorporated into the design and architecture of the data marts. The performance of the data warehouse had also improved with changes. The users were aware of efforts being undertaken to improve on the consistent look of applications across the products, to make them more user friendly.

The users were satisfied with the amount of training they had received on the data warehouse. They are able to understand the functions and features of the data warehouse and interpret its data adequately. The training provided to the users was informal and on an "as needed" basis. The users found it to be sufficient to carry out their functions. They found the data warehouse team very accommodating and received help from them whenever needed. The training was found to be adequate for the clients of the company to carry out their functions as well, because the clients were engaged very early on in the process of building the data warehouse.

Data Quality

At Nielsen Media Research, user participation, perceived usefulness of the data warehouse, ease of use, and high data quality led to greater user satisfaction and user acceptance of the data warehouse. A basic requirement for a successful data warehouse is its ability to provide business users with accurate, consolidated, and timely information. The accuracy, consistency, reliability, and timeliness of the data warehouse information were significantly high at Nielsen Media Research.

The users were satisfied with the accuracy of the data provided in the data warehouse. Because this data was the "currency" for the business, it was held to a higher standard. The users found the data to be very consistent and reliable as well. For example, data from 25,000 homes was input between 3:00 a.m. and 4:00 a.m. every day; after processing, this data was loaded into the data warehouse and was released to the clients consistently, starting at 7:00 A.M. Numerous checks and balances were incorporated into the quality assurance process to ensure the accuracy and reliability of the data. The data warehouse project leader and the database administrator made every effort to ensure the data's consistency and reliability.

The data warehouse provided timely information for all users and has always met the service-level agreements. The data warehouse is available 24 hours, 7 days a week to its users. It is available even during backup because the features in Sybase IQ allow reader nodes to access the database even while the writer node updates the database. Sybase IQ switches between the nodes automatically.

The data warehouse was perceived as useful and aligned to user needs at Nielsen Media Research. The data warehouse supported business users by integrating data from multiple, incompatible systems into a consolidated database. Users found the data warehouse easy to use and the data warehouse functions and features easy to understand. They were satisfied with the data warehouse team support and the training they had received on the data warehouse. Numerous checks and balances incorporated into the quality assurance process at Nielsen Media Research ensured the accuracy, reliability, and consistency of the data. The data warehouse provided information in a timely manner, meeting service-level agreements. All of this led to a successful adoption and use of the data warehouse to further the strategic initiatives of Nielsen Media Research. User satisfaction is important to achieve successful adoption of the data warehouse and its alignment with business strategies.

Technical Integration and Data Warehouse Alignment and Success

A number of factors contribute to and facilitate technical integration of the data warehouse. The alignment mechanisms of (a) value management, (b) technological capacity, and (c) organizational capability help in achieving technical integration in data warehouse projects.

Value Management

Value management is the organizational mechanism for ensuring that data warehouse resources invested throughout the organization deliver anticipated or greater returns (Henderson et al., 1996). In a data warehouse, technology should be selected based on its ability to address business and user requirements. Generally, making investments in technology does not in itself ensure profitability. In the data warehouse project at Nielsen Media Research, technology is evaluated after the business problem that it can help solve has been identified, saving time and resources and allowing the organization to focus on developing business solutions, and not just technology architectures.

At Nielsen Media Research, the data warehouse was built after careful technical evaluation consisting of both technical criteria and business metrics. The data warehouse architecture was selected after an in-house architect researched and explored the different options. Different data warehouse DBMSs were considered and vendor evaluations were completed. Each of the vendors were quizzed on technical and business metrics and weighted values were given to each criterion. Input was also received from Gartner Group, and proof of concept was done for the products evaluated. Sybase IQ was finally selected over NCR Teradata and IBM DB2.

The technology for the data warehouse was selected after comprehensive appreciation of business needs. As discussed previously, it was chosen after careful evaluation of technical and business criteria. The technology selected was appropriate to address business and user requirements. It had acceptance at the user interface, due to ease of use and perceived usefulness. It was also found that the users were satisfied with the training provided to them. According to the CIO, the technology invested had delivered more than anticipated returns, because the data warehouse had exceeded its expectation with respect to return on revenue.

The data warehouse was evaluated both through user feedback and through monitoring by the data warehouse development team using monitoring tools for the hardware and CPU. The loads into the data warehouse were continuously monitored and its performance evaluated. Because the output from the data warehouse affected the revenues of the company directly, feedback from the user was generally quick. The data warehouse had met the users' service-level agreements on all occasions.

Technological Capability

Technological capability deals with the administrative process for creating the required IT capability for supporting and shaping the business strategy (Henderson et al., 1996; Pollalis, 2003). In the data warehouse environment, technological capability involves integration of data from other systems and integration of data warehouse architecture into existing IT architecture. A high degree of technical integration of the data warehouse at Nielsen Media Research led to better strategic alignment of the data warehouse and a successful adoption of the data warehouse.

Integration of Data from Other Systems

At Nielsen Media Research the data was well integrated from different systems across the organization into the data warehouse. The data was integrated from mainframe, UNIX, and other sources into the staging area of the data warehouse. Data was integrated from the ETL process and the reference hubs into target tables that were then committed to the data warehouse tables. The integration of data from different systems in the organization into the data warehouse was not an easy process. The element of time presented the biggest challenge in the integration process, due to differing time zones and different timings of the source data.

Integration of Data Warehouse Architecture into Existing IT Architecture

The data warehouse architecture was closely tied to existing IT infrastructure and was fairly transparent to source systems. Data was received into the data warehouse from existing IT systems and data was passed from the data warehouse to these IT systems. The data warehouse intercepted the data from source systems and cleansed it. It integrated this data back into the legacy systems that were not yet a part of the data warehouse. Attempts are made to conform to naming standards. Nielsen Media Research is in the process of streamlining the different variations of the products and repurposing them due to aging technology. The data warehouse is the precursor to the efforts to service the products in a new way.

Organizational Capability

Organizational capability deals with the administrative processes for creating the required human skills and the capability for supporting and shaping the business strategy (Hwang et al., 2004). The skills of the data warehousing development team have a major influence on the outcome of the project and affect the adoption of data warehouse technology. Slow response to user requests or lack of resources and insufficient team members can lead to user dissatisfaction, which can affect the adoption of the warehouse.

At Nielsen Media Research, the business users, data warehouse managers, and senior managers recognized the importance of a highly skilled and well-equipped data warehouse development team and were satisfied with them. The users found the data warehouse team to be very skilled, highly motivated, and talented. They found the data warehouse team to be very responsive to their needs. Senior management was highly satisfied with the data warehouse team. They found the team to be knowledgeable in business and technology and felt that they had the right people with the right skill set. Steps are under way to offload the production and operational type of work from the data warehouse team to an operational department. This move was on the business plan because it was deemed expensive to use the data warehouse resources for operational processes.

Flexibility in Data Warehouse Planning

Lack of flexibility in data warehouse planning can impact the strategic alignment of the data warehouse and its successful adoption. Senior business managers, data warehouse managers, and the users of the data warehouse can have different perspectives on the impact of flexibility on the data warehouse framework and in data warehouse planning. The senior management perspective is influenced by changes that take place in business plans and the response of the data warehouse to changing needs. The data warehouse manager's perspective is influenced by the degree of flexibility in the data warehouse architecture and planning. The business user's perception of flexibility of the data warehouse depends on how well it responds to a change in business needs.

At Nielsen Media Research, major changes had taken place in the last 5 years in their business plans because of a dramatic increase in complexity in the industry. Even though the same business model had existed for a long time, and the overall business plans had remained the same, technological advances had introduced new methodologies for measuring television audiences. A major shift in business plans was the addition of the Local People Meter to collect household information and person-by-person data every day instead of on a quarterly basis. Another change was the introduction of new television technology in the form of DVR and VOD services in the market and corresponding business plans to measure audience viewings on them.

Senior business managers at Nielsen Media Research felt that the data warehouse responded well to changing needs. The move from diaries to the Local People Meter to capture the demographics of the market had caused changes in the Nielsen Media Research environment. The data warehouse had been developed as a result of this. The senior managers found the data warehouse response to be flexible to frequent reprioritization of business needs, as client needs changed. The data warehouse also allowed the products to develop faster. Strategically, the data warehouse is being leveraged to cater to modernization and cost-effective programs, though it was originally built to generate greater revenue.

The data warehouse accommodated new business needs and changes, because its existing architecture was extremely flexible. It had been built with flexibility in mind. The data warehouse has embraced re-factoring and continuous delivery as well as addition of new data types such as time-shifting data. The data warehouse allowed data structures to be extended efficiently to handle changes. Changes made to the base structure of the data warehouse have rippled through to products, with minimum downtime and effect. If the data warehouse were not flexible enough to react quickly to changing conditions, its usefulness would decline. New events remain a priority for the data warehouse. All the data warehouse managers agreed that the database, application middleware, and front-end tools had high scalability. In one application, Stellar, the front end faced some issues with scalability due to growth in its user base.

Data warehouse users found it responded quickly and thoroughly to changes in business needs. This was attributed to the experience of the data warehouse team members. They were experienced in both data knowledge and the technology involved. Business managers met regularly with the data warehouse managers at strategy group meetings, where business needs were conveyed to the data warehouse manager. Because the data warehouse team usually had a lot of business requests, they had to prioritize these requests. Depending on the severity of the request, some requests were met sooner and some later.

Alignment is seen as a process requiring continuous adaptation and coordination of plans and goals. A flexible data warehouse infrastructure makes it easier to introduce change in cost- and time-efficient ways. At Nielsen Media Research everyone views the data warehouse as being flexible to accommodate changing business needs. There does not seem to be a wide gap between emerging strategic direction and the data warehouse's ability to support it. Quick, iterative developments of the data warehouse have addressed the shifting of business requirements in a satisfactory manner. The flexibility of the data warehouse at Nielsen Media Research has allowed the organization to react to changes in its environment (new technology in audience viewing) and maintain the advantage its information provides.

Conclusion

This chapter presented a case study at Nielsen Media Research. It provided insight on how to achieve alignment of the data warehouse to business strategy and plans in a real-world scenario and showed its impact on successful adoption of a data warehouse. The factors that facilitated strategic alignment of the data warehouse also influenced the successful adoption of the data warehouse. Strong management support, joint responsibility between business and data warehouse managers, alignment between business plan and data warehouse plan, flexibility in the data warehouse framework, business user satisfaction, and technical integration of the data warehouse all influence the strategic alignment and successful adoption of the data warehouse.

Chapter 10

Case Study: Strategic Alignment at Raymond James Financial

Introduction

This chapter presents the findings from a case study conducted at Raymond James Financial. This case study delves in depth into the complexities of a data warehouse and its alignment to business strategies. It seeks to explore *how* data warehouses are aligned to business strategies and plans in the real world. It studies a complex case within its institutional context. The case study recognizes the complexity of the project undertaken and assesses the data warehouse in its natural, unaltered setting. It explores in depth the complexities and processes of a data warehouse and its alignment to business strategies and goals.

The first section of the chapter presents the overview and background of the company. The second section describes the data warehouse implementation. The third section of this chapter presents the way strategic alignment between the business objectives and the data warehouse was achieved.

Overview of the Company

Raymond James Financial was established in 1962 and has been a public company since 1983. It is now one of the largest financial services firms in the United States, with 2,100 locations worldwide. Its stock is traded on the New York Stock Exchange (RJF) and its shares are currently owned by more than 13,000 individual and institutional investors. Raymond James Financial was selected by *Forbes* (10 January 2005) magazine as one of the 100 best-managed companies in the United States.

Raymond James Financial is a diversified financial services holding company. Its subsidiaries engage primarily in investment and financial planning, including securities and insurance brokerage, investment banking, asset management, banking and cash management, and trust services. Through its four investment firms — Raymond James & Associates, Raymond James Financial Services, Raymond James Ltd., and Raymond James Investment Services — the firm had more than 5,100 financial advisors in 2,200 locations throughout the United States, Canada, and abroad, providing service to more than one million individual and institutional accounts. In addition, total client assets were over $136 billion, of which more than $25 billion were managed by the firm's asset management subsidiaries.

Company Subsidiaries

- At the time of the study, Raymond James Financial Services was a subsidiary company and consisted of 3,809 independent contractor financial advisors in 1,565 offices in all 50 states.
- Raymond James & Associates was a wholly owned subsidiary and consisted of 861 employee financial advisors in 76 offices concentrated in the South, Mid-Atlantic, and Midwest.
- Raymond James Ltd. consisted of 259 financial advisors in 46 offices, all located in Canada.
- Raymond James Investment Services consisted of 53 independent contractor financial advisors in 25 offices. The company owns 75% of this joint venture in the United Kingdom.

Business Objectives

Raymond James Financial provides individual investors, corporations, and municipalities with investment, financial planning, investment banking, asset management, banking, and trust services throughout the United States, Canada, and internationally. Raymond James Financial had more than 5,000 financial advisors, both employees (RJA) and brokers (RJFS), who worked directly with the investor's clients.

In its mission statement, Raymond James Financial announces as its first precept that clients always come first. As a client-oriented company, it therefore aims to provide the highest level of service and have clear and frequent communications with its clients. It aims to maintain superior quality in its service and cooperate with and provide assistance and support to its associates. Its mission is to also provide continuing education and excel beyond its peers.

Corporate Structure

Raymond James Financial has a traditional (hierarchical) corporate structure. Its subsidiaries operate independently but compete collectively, focusing on a diverse range of businesses. It is headed by the president, who is also the CEO (chief executive officer). The COO (chief operating officer) and the presidents of all the subsidiaries report directly to him. The CIO (chief information officer) is a corporate officer at the vice presidential level and reports directly to the CEO. With only one level separating this person from the top of the organizational hierarchy, the CIO wields great decision-making authority and control over a large IT organization. The information technology division comprises four major functions: (1) software development and production management, (2) operations, (3) database administration, and (4) project management, which helps integrate all areas. The data warehouse falls under the software development and production management function. The chief data warehouse development officer reports directly to the CIO.

Background of the Data Warehouse

Raymond James Financial operates in a vast, complex, and competitive financial market. In an industry that has generally consolidated into a relatively few large conglomerates, Raymond James has continued to remain independent and to grow. To sustain this growth it becomes imperative for them to help build loyalty between clients and the Raymond James advisors.

The long-term operating plan of Raymond James Financial focuses on improving their return on equity (ROE) through an effective deployment of excess capital and a reengineering of the processes utilized in operations and information technology. The objective is to improve their effectiveness and efficiency, as well as their service to clients and financial advisors, by elevating the quality and productivity of their sales force.

The data warehouse provides an opportunity to improve their efficiency and serve the needs of the financial advisors better. The idea of building a data warehouse emerged in 2002. It grew out of the recognition that the financial advisors had to look in many different places for information and reports. The large inventory of information sources (more than 20) and different interfaces placed an additional burden on the financial advisors to first become aware of all of these sources and then access multiple user interfaces (i.e., applications, reports, and websites) to retrieve information. Moreover, in providing information, these sources oftentimes overlapped with each other.

In addition, having similar information stored in multiple databases provided a greater probability that the information would differ across these databases. For example, many times the retail division as well as other divisions within Raymond James created their own local databases to provide information not currently provided by existing data sources and their user interfaces. Loading and maintaining all of these databases significantly increased the cost of doing business.

A need was also felt for more detailed and relevant reports. Certain reports core to the Raymond James business had been in place for a significant period of time and had not kept up with the changing business environment and technological advances. Some paper-based reports provided only a static view of key information, which did not answer all the key business questions. Again, some other reports were distributed on a weekly or monthly basis and provided summary information and not the underlying detail. A need was felt for more immediate online access to information as well as the ability to provide different views and reports of this information based on the specific needs of the financial advisor or management.

Data Warehouse Objective

The objective of the data warehouse was to provide the financial advisors, financial advisors' management, and other Raymond James Financial individuals and divisions with one data warehouse that provides current, dynamic access to consistent and accurate information on key performance indicators for the business. This data warehouse would provide the ability to not only report information but also leverage this information to increase profitability, both by reducing costs and expenses as well as by increasing production.

Data Warehouse Development

Conceptual Model

The first step in building the data warehouse was to provide a strong foundation of common information. An enterprise foundation was created on which subject areas were then built. This foundation comprised information on organization, products, account types, time, general ledger buckets, asset ranges, transaction types, production types, financial advisor types, and demographics. This common information provided all users with a single point of reference when accessing subject-specific data, such as account assets or a financial advisor's production numbers.

Subject areas for assets, production, and account statistics were then created on top of this foundation. These subject areas provided a flexible, dynamic online view, which replaced and enhanced the information that was being provided. These subject areas are available not only to the financial advisor community but also to other Raymond James employees and divisions.

Physical Implementation

The data warehouse was built on an HP Superdome platform with 12 processors, 48 gigabytes of memory, and a 64-bit MS Sequel server, Enterprise Edition. It sourced data from 20 different databases into a staging relational database of 10 terabytes. Data was cleansed, integrated, transformed, and then loaded into the three data marts. It used online analytical processing from Microsoft (MS OLAP) to build the subject areas for reporting and analysis. OLAP is a database software that provides an interface, so users can transform raw data quickly, and interactively examine the results in various dimensions of the data. The MS OLAP rolled the three data mart views up to an enterprise view of the data warehouse. The data warehouse contains mainly of historical data up to 2 years old, and it is current up to the previous close of day. The front end of the warehouse comprised a custom application MDX (a Microsoft term for multidimensional expressions), a visualizing component, which directly accessed MS OLAP. It consisted of mainly pre-formatted reports. Data was accessed for analysis through Proclarity, a third-party analysis tool, chosen from a limited number of access tools that can service MS OLAP.

Project Initiation

The data warehouse project was initiated by the CIO, who was instrumental in presenting and selling the project to the business and CEO. It was one of the top five projects in the company at that time and enjoyed high visibility at the upper management level. The data warehouse is a joint effort between the business and IT division. The IT division, having heard the requests over the years, worked closely with business managers and financial advisors to determine the data warehouse project's impact on business and its priorities. The data warehouse was built to an initial cost of $3 million, which the business divisions invested into the data warehouse as a shared business expense. The CIO was involved in all higher level decisions and major expense decisions, for example, the purchase of major hardware, large software expenses, or new human resource expenses.

Current Data Warehouse

The data warehouse project has been in existence for more than 2 years. A successful prototype was built in the first 8 months as proof of concept. The prototype was then enlarged over the next 1½ years into the current data warehouse. An iterative development methodology was adopted to build the data warehouse. This enabled the demonstration of the progress of the data warehouse to the financial advisors every 3–4 months at regular conferences. These conferences allowed the articulation of requests and information needs from the management and financial advisors and their inclusion in the data warehouse development process.

The data warehouse is available for use by up to 7,000 individuals. Every financial advisor has the ability to use some form of data from the data warehouse. Different levels of users have been assigned different levels of access to the data warehouse, depending on the security required. Users have been trained using online sessions and Web-based training for the data warehouse. Data warehouse classes are held for financial advisors, who then receive a certificate at the completion of their training. Around 500 financial advisors and managers use it every day for reporting and analysis. The activity level on the data warehouse changes over time. Activity level is at its highest on Mondays and near month's end.

Strategic Alignment of the Data Warehouse at Raymond James Financial

This case study demonstrates the impact of strategic alignment of the data warehouse to business strategies and plans on its successful adoption. It shows that strategic alignment between the business and data warehouse can be achieved by ensuring that the underlying critical factors — joint responsibility between business and data warehouse managers, alignment between business plan and data warehouse plan, flexibility in the data warehouse framework, business user satisfaction, and technical integration of the data warehouse — are addressed.

Joint Responsibility Between Business and Data Warehouse Managers

Commitment and Involvement of Senior Management in Data Warehouse Project

At Raymond James Financial, the CEO and the vice president of software development along with the senior business managers and product managers are very involved in data warehouse investment decisions. The CIO acted as the champion of the data warehouse project and was responsible for initiating and selling the data warehouse to the business. Strong support at the senior levels of management is important to data warehousing projects. This support and commitment by senior management has resulted in active promotion of the project (the data warehousing project was identified as one of the five most important projects in the organization) and ensured availability of capital, human support, and internal resources during the development and implementation process.

At Raymond James Financial, the senior managers were involved in the data warehouse investment decisions and were responsible for funding the data warehouse and getting resources from the operating committee of the company and the business units, who have ownership of the data warehouse. Although the CIO was the champion of the data warehouse project, the CEO participated in investment decisions involving more than $1 million. The CIO was instrumental in selling the data warehouse project to upper management, making it visible among senior management, and finding support for it.

Strategic alignment of the data warehouse depends on a strong relationship between the data warehouse managers and the business executives. For the enterprise data warehouse to be successful, a joint partnership is required between the business and IT managers. At Raymond James Financial, the VP of software development reported directly to the CIO heading a group of 330 people and worked closely with the heads of product management and project management. Senior managers were involved in sizing the data warehouse team and allocating its resources. The product management group, along with the product manager for the data warehouse, made decisions about moving people to the data warehouse and other teams, sizing the teams, and allocating resources. Decisions about moving forward in a project and funding resources were made along with IT operations and IT engineering groups.

The senior managers participated in defining the deliverables of the system as well as its display layouts. The product management group was involved in data warehouse investment decisions and met with the CEO and COO for presentations and funding. The product manager was involved in building a long-term roadmap for the data warehouse. He was also responsible for getting together the resources from the operating committee of the company and the business units who have the ownership.

Involvement of Data Warehouse Managers in Corporate Strategy

At Raymond James Financial, there were differing levels of involvement of data warehouse managers in corporate strategy. The upper management involved them in task forces dealing with new initiatives to improve current processes and to improve financial advisor relationships. Corporate strategy decisions were made at the CIO and VP levels of the company. Data warehouse managers were aware of major organizational business plans linked to the data warehouse applications in marketing and compliance, performance reporting, and contact management.

There had been a shift in corporate strategy decisions at Raymond James Financial. The corporate goal had shifted to acquiring financial advisors and the assets from under their management into the company. There had also been a focus on the use of technology as being a differentiating factor for recruiting more financial advisors. *The data warehouse was thus being used as a strategic and competitive tool.* The change in focus had led to the changes in the organization's IT plans. Efforts were under way to improve processes, and new software development methods were being rolled out. Also, within the company there was an ensuing discussion and analysis of the buy-vs.-build option, and focus was moving toward buying third-party products.

However, a difference existed in the level of awareness of corporate strategy by data warehouse managers and their involvement in corporate strategy decisions. The level of awareness of corporate strategies by data warehouse managers was high, but the level of involvement of the data warehouse managers in corporate strategy making was low. This status is not conducive to the strategic alignment of data warehouses. Joint involvement of data warehouse managers and senior managers is required to ensure strategic alignment of the data warehouse to business plans.

At Raymond James Financial, this problem is being addressed. A recent organizational change in the management structure of the company underlines the importance of the understanding and joint responsibility between data warehouse managers and business managers. Until now, the data warehouse managers had been participating in meetings with the business users and business managers. Now, a new layer in the form of product manager has been added to the data warehouse management. The product manager has more than 5 years of extensive business experience at the company and has been given the task and role of managing the data warehouse.

The product manager is now responsible for building a long-term roadmap for the data warehouse, participating in defining the deliverables of the system, as well as being responsible for getting resources for the data warehouse from the operating committee of the company and the business units who have the ownership of the data warehouse. The product manager works with the business managers on all data warehouse issues and involves the CIO in higher level decisions regarding the data warehouse.

The product manager is thus in the unique organizational position to set the stage for alignment of the data warehouse in its subsequent implementations. The understanding, commitment, and involvement of the product manager in the data warehousing process would have an effect on improving the strategic alignment of the data warehouse to business strategy in the future.

Alignment between business strategy and data warehouse strategy requires a strong interdependent relationship between the business and data warehouse managers. At Raymond James Financial, the product managers, data warehouse managers, and business managers are jointly trying to align the data warehouse, addressing deficiencies that have been observed. A high level of involvement and commitment of senior managers and data warehouse managers is critical to successful alignment and adoption of the data warehouse.

Alignment between Business and Data Warehouse Plans

Data warehousing strategy is impacted by the business strategy of the organization and impacts the business strategy in return. Alignment between the two — data warehousing and business strategy — is achieved by integrating the data warehouse and business plans. Aligning the data warehouse plans to the business plans creates synergy and adds value to the business. A number of factors contribute to, and facilitate, alignment between business and data warehouse plans. Hence, to gain a comprehensive understanding of the degree of strategic alignment, this section considers (a) architectural alignment of the data warehouse, (b) knowledge sharing, (c) integration of business and data warehouse planning, and (d) communication between business and data warehouse managers.

Architectural Alignment of the Data Warehouse

Alignment of the data warehouse strategy and architecture to business strategy and architecture ensures its successful adoption. Misalignment between IT architecture and business architecture could mean higher costs and a loss of opportunities.

There was a need at Raymond James Financial to collect and integrate data from assets and revenue, together with historical and statistical data, and analyze it as a whole. The data warehouse architecture was selected and built in response to requests made by business divisions and financial advisors. The objective of the data warehouse project was to structure into one system all the information needed for the financial advisors and the management. Toward this goal, three data marts were formed for the three core divisions: assets, revenues, and accounting. These data marts rolled up into an enterprise-wide data warehouse. The enterprise-wide data warehouse was selected to enable financial advisors to see all aspects of the business.

At Raymond James Financial, the access to the data warehouse was one of the key competitive advantages the company had been able to offer in recruitment of financial advisors. Additionally, the data warehouse also supported the analytical functions in the business. For example, the data warehouse had been helpful in analyzing the impact of the imposition of or raise in a particular fee on the accounts held and how the fee could be instituted. However, the very complexity that made the data warehouse powerful was making it hard to use. A team of users with greater expertise was required to fully exploit and use the data warehouse.

The data warehouse supported organizational business plans at different levels to different capacities at Raymond James Financial. Initially there had been growing pains in the architectural alignment of the data warehouse, and the amount of data in the data warehouse seemed overwhelming. As the familiarity with the data warehouse increased and its use increased, the financial advisors found it to be of immense help in analyzing and planning their business. This offered them a significant advantage, because before the data warehouse was developed, the financial advisors did not have access to integrated data and information was mainly extracted from scattered reports.

For areas where the data warehouse did not support the business plans well, it was attributable to the fact that different functional groups within the company functioned differently. There was a need to share appropriate data within the functional groups, and this was not adequately reflected by the data warehouse. There were ownership issues with data within the functional groups. Management was aware of the potential of the data warehouse to realize the vision of the company. There was a perception held that the data warehouse could help in diminishing these boundaries and facilitate sharing of information. The data warehouse could help in the de-segregation of data by integrating it.

The data warehouse project was initiated 2 years before the time of the study and went live a year before the time of the study. Since going live, the plans have been scaled back. There was more focus on the initial rollout of the business analyzer. Over the past year, there have been several user requests to analyze institutional data through the data warehouse. Business users have made requests to expand the data warehouse beyond the three core subject areas to other areas such as expenses and bank data. However, the data warehouse plans have been scaled back due to resource constraints and institutional priorities, and there has been less focus on its use across the organization by adding additional subject areas. Even though the data warehouse could not satisfy all current user requests due to resource constraints, the data warehouse managers realized the need for more subject areas.

Initially, the scope of the business vision dictated the architectural approach. But currently, the organizational vision and strategic plans do not provide a roadmap for the data warehousing effort. A strategic objective of long-term gain and full organizational control is necessary for an enterprise data warehouse architecture. The data warehouse managers and senior management are aware of this, and steps are being taken to address this disconnect between strategic vision and data warehouse plans. Efforts are under way to address this deficiency and align the data warehouse architecture to business strategy and architecture.

Knowledge Sharing

Knowledge sharing between the business and data warehouse managers improves alignment between data warehouse and business plans. Knowledge sharing avoids decision making that is out of alignment and can help to integrate the business and data warehouse planning process.

At Raymond James Financial, the data warehouse managers were receptive to business managers' needs and vision. The data warehouse managers at Raymond James Financial had helped the business managers understand the advantages and limitations of the data warehouse technology and were in turn aware of the business plans and strategies. The data warehouse managers had communicated to senior managers that the data warehouse was not flexible enough to adapt to all their current needs and that they were struggling with it. The data warehouse managers had also informed them of the need for more resources and funds required for the incremental releases of the data warehouse. The senior managers understood the necessity of these resources but were finding it hard to justify them financially.

Integration of Business and Data Warehouse Planning

Integration between business and data warehouse plans is critical to strategic alignment and successful adoption of the data warehouse. At Raymond James Financial the data warehouse planning process was not well aligned with the business visions and plans. This situation was leading to poor alignment.

Initially, in the first 2 years, the data warehouse was aligned to business strategies. But following changes to the business strategies and business demands, the data warehouse's speed of change, to meet these changes in strategic direction and demands, was slow. The filtering down of business plans through the management layers was also slow. The data warehouse team had also met with resource constraints. Issues had arisen at Raymond James Financial in integration of business and the data warehouse planning process. To address these problems, the product manager had been added to the management structure of the data warehouse. Integration of business and data warehouse planning was now being facilitated by the product manager, who was involved on both sides.

The product manager was responsible for integrating the data warehouse with business needs and business strategy. The product manager understood the business side of the company and had more than 5 years of experience in it. The position of the product manager was added to the management structure of the data warehouse to enhance the cooperation and coordination between the data warehouse and the business. Senior managers felt that this change in the management structure had shown signs of being effective and they received good feedback from the business. The needs of the different groups within the business were different. For example, the needs of the private client group and those of the capital groups were quite different. Because of this, the product manager prioritized the user needs and determined costs and ways to drive future releases of the data warehouse.

Integration of data warehouse plans with business plans is important to its strategic alignment. Poor alignment in costs, both time and resources, and does not deliver expected results. At Raymond James Financial, the data warehouse plans were developed in response to business plans. The business needs and business strategy were used to define the data warehouse architecture and processes and led to its strategic alignment and successful launch. But later the divergence among the data warehouse plans, business needs, and business plans led to suboptimal outcomes.

The senior management and the data warehouse managers at Raymond James Financial are aware of the problem and they recognize the importance and necessity of integrating business and data warehousing plans. Alignment can be achieved by partnering the data warehouse leadership with the business leadership and engaging the business in the planning processes of the data warehouse through open communications. The introduction of the product manager is a step in this direction to enhance the cooperation and coordination between the business and data warehouse plans and positive action on the part of Raymond James Financial to integrate data warehouse plans with business needs and plans.

Communication between the Business and Data Warehouse Managers

Communication between all stakeholders of the data warehouse is essential for its alignment to business strategy and adoption by the business users. Along with knowledge sharing and communication of the strategic direction to the data warehouse managers by the senior management, creating frequent communication channels between data warehouse managers and users is necessary to facilitate better understanding of the data warehouse and its ultimate successful adoption.

At Raymond James Financial, both formal and informal communication channels existed within the organization. One of the formal channels of communication between the business and data warehouse manager flowed through the product manager. The product manager was a fundamental and open line between the two sides and allowed free and open communication. The product manager validated the various requests coming through the different business divisions and conveyed them to the data warehouse managers.

The change management process used within the IT division was another formal medium of communication. But because the change management process consisted of completing several IT forms which were difficult to fill out, it was occasionally circumvented, and the business managers talked directly to the data warehouse managers. Another communication channel was the weekly staff meetings held by the vice president of software development, in which the product manager and data warehouse managers sometimes participated.

Formal and informal communication channels existed between the users and data warehouse team. Where there were no formal or established communication channels between the users and the data warehouse team, the communication between the user and the data warehouse managers and team was informal. Generally the user communicated to the warehouse team via emails or through phone calls. Sometimes ad hoc meetings were held.

Users also communicated their needs through the learning specialist to the data warehouse managers. In the initial stages of the first implementation of the data warehouse, users communicated directly with the data warehouse team. They had committees initially and communicated through email groups and intranet test sites. However, now the users communicated to the data warehouse managers via the training team and the learning specialist.

The learning specialist met with the data warehouse developers and managers for an hour every Tuesday, in a face-to-face meeting. In this forum, the user issues were communicated to the data warehouse team, but usually there was no resolution to these issues. The outcome usually was a compounding list of user needs communicated to the warehouse team. The users were also supposed to communicate directly to the product managers. However, the decisions of the product manager or data warehouse manager were not communicated back to them. Some users received little or no feedback from the data warehouse managers or product manager. Users were not highly satisfied with the communication in the opposite direction, by the data warehouse team to the users. Overall, the communication between the business managers and the data warehouse managers seems to flow, either directly or with the facilitation of the product manager.

There was a lack of good two-way communication between the data warehouse managers and its users. Although the user needs are communicated either directly or through the learning specialist to the data warehouse managers, a reciprocal feedback from data warehouse managers to the user was not satisfactory. This is an area of communication that can be improved for better adoption and use of the data warehouse. The observation by a user of the need for more effective communication and knowledge sharing among the various IT divisions and the data warehouse team, to keep information in the data warehouse consistent, further highlights the importance of good communication for better alignment of data warehouse and business plans. Lack of communication has led to user dissatisfaction.

Business User Participation and Satisfaction, and Data Warehouse Success

A number of factors contribute to the degree of participation and satisfaction of the data warehouse user. To gain a comprehensive understanding of the degree of user participation and satisfaction, this section considers the following: (a) user participation, (b) perceived usefulness, (c) ease of use, and (d) data quality.

User Participation

At Raymond James Financial, not all users were involved in the development of the data warehouse. A few users were involved in defining the user specifications. Other users were involved in testing reports that came out of the data warehouse after implementation to validate output. Some other users were involved in defining the report. Yet other users were asked to contribute suggestions on how to train the users. A majority of the users participated as users of the data in the data warehouse only. Users were not very satisfied with their participation in the overall data warehouse project. It was observed that the people who ultimately used the data from the data warehouse were not adequately involved in the design of the data warehouse. The users wanted greater involvement and participation in defining user requirements, report specifications, and design of the queries. User involvement during user requirement definition can significantly impact user satisfaction. The users at Raymond James Financial experienced limitations in the use of the data warehouse due to being unaware of the data design, data storage, and accessibility to the data in the data warehouse.

Participation by the users in the data warehouse project is important. End-user participation has a direct impact on the adoption of the data warehouse. The selection and inclusion of appropriate users in the project team is essential. The users of the data warehouse at Raymond James Financial were not satisfied with their level of participation in the development of the data warehouse. User participation is essential for better communication and coordination of user needs. End-user participation is helpful in managing user expectations and satisfying user requirements.

Perceived Usefulness

For the data warehouse to be accepted and be successfully adopted by the users, it should be perceived as useful by its users. At Raymond James Financial, because of the integrated data available in the data warehouse, an individual analyst did not have to make several points of contact to find information. It had made access to information more organized and centralized. Yet the business users of the data warehouse at Raymond James Financial were only moderately satisfied with the relevance of the data warehouse to their day-to-day decisions. The data warehouse did not adequately support the users' needs. The users felt that the data warehouse could be better used and had greater potential.

Not all the users of the data warehouse at Raymond James Financial perceived that the data warehouse was useful to their needs. The data warehouse was used by the users on a weekly or monthly basis for analytical purposes and business planning. It was used primarily for querying, reporting, and looking up historical data. Some users wanted to use it more frequently and felt that it could be used for day-to-day decisions. But they were unable to use it adequately because of the difficulty encountered in finding the desired information in the data warehouse, the non-intuitive design of the user interface, the data warehouse "locking up" while running reports, and drawbacks in preparing reports. Occasionally, users resorted to other information systems that existed within the company to satisfy these information needs.

The users observed that although the platform of the data warehouse had all the information they sought, its design made it difficult to extract the necessary data. For example, financial advisors could view their individual assets but could not view assets at the corporate level. The users found it difficult to merge data from the data warehouse with other files or external data because the data warehouse information lacked unique identifiers.

Of the users that found the data warehouse supported their needs, it was the availability of integrated data in the data warehouse that they found the most useful. Before the data warehouse, they had limited ability to collect, integrate, and analyze data. They also found the available canned reports useful.

The strength of the data warehouse lies in its ability to organize and deliver data in support of management's decision-making process. The response of senior managers at Raymond James Financial to the usefulness of the data warehouse varied from its business users' perception. Senior management perceived the data warehouse as useful. They usually felt that the data warehouse provided them with the information they needed. The data warehouse was being used to easily see the correlation between the account side and revenue. It was being used to drive behavior from sales management. Some managers believed that even though the data warehouse did not provide them with the information they sought, the data warehouse was designed for the financial advisors and it was they who would find it useful. For example, the data warehouse provided information on the viability of a particular fee, and applying this fee made a direct impact on the profitability of the company. Some senior managers found that the data warehouse had also resulted in a new business strategy. The data warehouse had been used by management to monitor the progress of its new model for recruitment of financial advisors. Senior management found the access to organized and centralized information in the data warehouse useful. For instance, some changes had been facilitated by the data warehouse. Departments had made policy changes and had enforced these policies based on data from the data warehouse.

Ease of Use

Business user satisfaction is also dependent on the user experience of the data warehouse. The users at Raymond James Financial did not find the data warehouse easy to use. It was observed that the data warehouse was not intuitive to use and that lack of user manuals made it even harder to use. The financial advisors found the data warehouse difficult to navigate and time consuming. This difficulty in using it had prevented users from exploring all the options provided in the data warehouse and was therefore resulting in its infrequent use. Some users felt that greater technical expertise was required to fully utilize the data warehouse. In fact, many options available for use have not yet been touched upon by users. The users found the data warehouse interface highly technical, requiring users to have previous knowledge of how and what to do in order to use it. The data warehouse was not found to be intuitive and lacked user manuals for reference. The slow speed of the queries was another deterrent.

Besides query speed, the users faced difficulty in knowing which cube the information sought lay in and how to extract it. They found the data warehouse to be complex, time consuming, and lacking a user's guide. The users observed that querying for information on the data warehouse was not "straightforward." For example, data could be sliced by region but not a combination of regions. The users would like to see additional ways to slice the data in the data warehouse. Other users found it difficult to understand the logic of sorting data in the data warehouse. The data warehouse would often "hang" the computer while running a query it could not accommodate. Overall, the users found the data warehouse functions and technical features difficult to understand. They felt that a background in a database tool seemed necessary to fully understand and use the data warehouse.

At Raymond James the users were dissatisfied with the ease of use of the data warehouse. The users were dissatisfied with their understanding of the data warehouse functions and features. The users were also not satisfied with the amount of training they received on the data warehouse. The data warehouse should not simply be a very large database. It should be able to provide information necessary to answer business questions in a manner that is comfortable and intuitive to the business user.

In the data warehousing environment, the business users are the main customers of the system. A basic requirement for a successful data warehouse is its ability to provide its users consolidated information in an intuitive and comfortable manner. The competitive advantage of a data warehouse depends on the bulk of the organization's employees being able to quickly and easily access the data and interpret the information.

Data Quality

Data quality is pertinent to the success of the data warehouse. The issues of data quality in the data warehouse can be characterized into accuracy of data, consistency and reliability of information, and timeliness of data in the data warehouse. Problems in any of these areas of data affect the quality of data in the data warehouse.

At Raymond James Financial, not all users were satisfied with the accuracy of the data provided in the data warehouse. Although the data warehouse contained mostly accurate data, complete accuracy was sometimes questionable because it had discrepancies with data from older systems. Once the source of the data was obtained, the discrepancies were sorted out. In some cases, data was not being drawn into the data warehouse from the right source. In other instances the query had inclusion or exclusion criteria that were not obvious to the user and the data the user received may not have been what the user had wanted.

Some users found that the reports drawn from the data warehouse sometimes did not appear to be accurate. The discrepancies in the reports may not necessarily be because of problems in the data warehouse. They may occur (a) because of the way data was sourced into the data warehouse, or (b) because the data warehouse reports may contain more data than similar reports made from other sources, resulting in disparate figures. These discrepancies are important because a minute difference in detail between two similar reports can often translate to significant differences in dollar terms. These discrepancies led to reduced acceptance of the data warehouse. When users encountered these inaccuracies early on in their use of the data warehouse, they did not want to use the data warehouse again. Again, inaccuracies found in one area of the data warehouse kept the users from using other areas of the data warehouse. For example, inaccuracies in the new accounts system led users to expect inaccuracies in the business analyzer as well.

The data warehouse provided mostly consistent data (information was available on a monthly basis), but it was not always reliable. The data in the data warehouse went back only to the beginning of 2005, hence occasionally the data had to be cross-checked with reports from other systems to get a more complete picture. Sometimes, the data in the data warehouse was consistent but the complexity of use was such that different queries run to find similar information resulted in different answers. The user confidence in the data was not always complete.

The data warehouse provided timely information for some users, but for some other users, it did not. The data was updated in the data warehouse on a nightly basis and reflected data from the close of the previous day. Users could physically get the data from elsewhere but had to wait a day to get it from the data warehouse. It therefore affected the efficiency of the user. Additionally, the manipulation of data to satisfy some queries took time, affecting user efficiency. Providing timely information was an area of opportunity for the data warehouse at Raymond James Financial. There was room for significant improvement in areas concerning accuracy, reliability, and timeliness of the data in the data warehouse. The introduction of metadata into the data warehouse could address the data integrity issue to some extent.

A basic requirement for a successful data warehouse is its ability to provide business users with accurate, consolidated, and timely information. Failure to gain credibility with the business users may lead to failure in aligning the data warehouse to business strategies. The degree of business user participation and satisfaction is critical to strategic alignment and successful adoption of the data warehouse. At Raymond James Financial, difficulties encountered in using the data warehouse had created dissatisfaction among the business users. The users were dissatisfied with the ease of use and the data quality of the data warehouse. Dissatisfaction also existed among the business users, with their low level of participation in the development of the data warehouse. All these factors — user participation, perceived usefulness, ease of use, and data quality — are important to achieve user satisfaction. User satisfaction is important to achieve successful adoption of the data warehouse and its alignment to business strategies.

At Raymond James Financial the data warehouse managers are aware of the dissatisfaction among the users and the importance of user satisfaction for the successful adoption of the data warehouse. Efforts are under way to address the current needs of the users.

Technical Integration and Data Warehouse Alignment and Success

A number of factors contribute to and facilitate technical integration of the data warehouse. The alignment mechanisms of (a) value management, (b) technological capacity, and (c) organizational capability help in achieving technical integration in data warehouse projects.

Value Management

Value management is the organizational mechanism for ensuring that data warehouse resources invested throughout the organization deliver anticipated or greater returns. In a data warehouse, technology should be selected based on its ability to address business and user requirements. At Raymond James Financial, new technology at the user interface had low acceptance due to low perceived ease of use and low perceived usefulness. It seems that despite the potential benefits of data warehousing, the tools provided to the users cannot be used easily, resulting in non-utilization of the tools, and therefore reducing return on investment.

Users dissatisfied with the training provided to them also contributed to low value management. The level of training in users was varied, and not all users found the training given to be adequate. One user received 2 days of training for the desktop version and Web browser of the data warehouse; another received only 2 hours of an overview of the data warehouse and canned reports. The users thought the training provided on the data warehouse was basic. They felt that more directed training was required to meet their needs.

In data warehouse projects, evaluating technology after the business problem that it can help solve has been identified saves time and resources and allows organizations to focus on developing business solutions, and not just technology architectures. At Raymond James Financial, the data warehouse technology was evaluated after the decision to build the data warehouse. The data warehouse was evaluated both through user feedback and through monitoring by the data warehouse development team.

The data warehouse development team monitored the performance of the data warehouse by monitoring how long the system took to load, its OLAP processing speed, and the users' drilldown analyzing capabilities. The user feedback received by the data warehouse managers reflected the slow speed of the data warehouse and the need for more user training.

Technological Capability

Technological capability deals with the administrative process for creating the required IT capability for supporting and shaping the business strategy (Henderson et al., 1996; Pollalis, 2003). In the data warehouse environment, technological capability involves integration of data from other systems and integration of data warehouse architecture into existing IT architecture. Technological capability at Raymond James Financial involved addressing both (a) integration of data from other systems and (b) integration of data warehouse architecture into existing IT architecture.

Integration of Data from Other Systems

At Raymond James Financial, data was integrated from different systems across the organization into the data warehouse. The data warehouse was fed from a core CSS system, which is a source of records for accounts, customers, demographics, and product type information. The data warehouse also integrated data from 20 different databases maintained in the company. At a staging database, the data was cleansed and transformed before loading into the data marts. The data was available till previous close of day in the data warehouse.

The integration of data from different systems in the organization into the data warehouse was not always smooth, and issues were encountered in the process of keeping the data warehouse in sync with all the other systems. Changes in the data in other systems had to be accommodated into the data warehouse. Team meetings were held to communicate and keep abreast of these changes. Changes in data were noted in such meetings and target dates were set to align the data warehouse with the implementation dates of these changes.

Integration of Data Warehouse Architecture into Existing IT Architecture

The data warehouse architecture was well integrated into existing IT systems architecture at Raymond James Financial. Raymond James Financial was predominantly a Microsoft shop, and the data warehouse's Sequel server was a Microsoft solution. The data warehouse was integrated into the mainframe system in the company and replicated the mainframe data into the sequential environment. The data warehouse interfaced with other IT systems through portal applications. A page in the regular application interface was linked to the data warehouse. It was ensured that the data warehouse kept up with changes in the mainframe.

Strategic alignment between business and the data warehouse can have a positive impact if the data warehouse is part of a well-integrated organizational IT system. At Raymond James Financial, the data from other systems have been well integrated into the data warehouse, and the data warehouse architecture is well integrated into existing IT architecture.

Organizational Capability

Organizational capability deals with the administrative processes for creating the required human skills and the capability for supporting and shaping the business strategy (Hwang et al., 2004). The skills of the data warehousing development team have a major influence on the outcome of the project and affect the adoption of data warehouse technology. Slow response to user requests by the data warehouse team or lack of resources and insufficient team members can lead to user dissatisfaction, which can affect the adoption of the warehouse.

At Raymond James Financial, the users of the data warehouse were moderately satisfied with the data warehouse project team and its ability to manage and solve technical problems. Users were also moderately satisfied with the response of the data warehouse team to user needs. This was because although the data warehouse team members were very skilled, they were short-staffed. There were too few members on the data warehouse team to meet all user needs effectively and promptly.

Although they made a very good effort at solving technical problems, occasionally the responses took longer. The data warehouse supervisor was perceived as being very competent and helpful in resolving user problems. (In his absence, the users had to wait for days for the problem to be handled.) The learning specialist provided support. But a lack of resources coupled with lack of sufficient members on the team led to a generally slow response to its users. A well-staffed data warehouse team could respond to user needs more efficiently and contribute to enhanced user satisfaction.

The skills of the data warehousing development team have a major influence on the outcome of the project and affect the adoption of data warehouse technology. A highly skilled team is better equipped to manage and solve technical problems. The business users, data warehouse managers, and senior managers at Raymond James Financial are aware of the need for and recognize the importance of a highly skilled and well-equipped development team.

Flexibility in Data Warehouse Planning

The issue of flexibility in planning is pertinent to the data warehouse environment. In data warehouse projects, the business strategy could change during data warehouse development. A change in business plan could lead to a shift in the business requirements of the data warehouse. As business needs change over time, a data warehouse needs to be flexible enough to be responsive to them, both in its architectural alignment as well as relevance of the data to the user.

At Raymond James Financial, major changes had taken place in their business plans in the last 5 years. The business plans changed to respond to changes in the regulatory industry, to margin compression, to industry downturn after the 9/11 incident, and to corporate scandals. Over the past 5 years these changes led to an increase in the trade volume of the company. The current management focus is on building systems that would provide information on an unsuitable trade and on compliance within the company. The data warehouse was being used to address this focus.

Yet another major change had been in the way the financial advisors affiliated with a firm or did business in the brokerage industry. Over the past 2 years, there were changes in the business plans, and Raymond James Financial introduced a new and unique recruitment model to the industry, becoming the leader in its segment and generating a demand in the industry.

The data warehouse had to respond to these changing needs in the business environment. Business managers felt that the response of the data warehouse team had been good to the end users, but the team's response to back-office accounting and administration had not been as satisfactory.

Business plans in some segments of the company had not changed in the past 5 years, e.g., focus on maintaining the growth and integrity of advisor relationships. The business plans require provision of best tools to the financial advisors to service their clients. The data warehouse was being used to implement this business plan, by supporting the financial advisors. The data warehouse architecture had not changed in the past years. The data warehouse database, application middleware, and front-end tools had high scalability because the data warehouse was built on an HP Superdome with 12 processors, 48 gigabytes of memory, and a 64-bit server along with a Web front-end.

To accommodate new business needs and changes, the existing architecture of the data warehouse was quite flexible and new dimensions could be added to it fairly easily. Recently, changes had been made to the data warehouse's existing dimensions and new levels of hierarchy were added. An example of the data warehouse's flexibility: it was observed that 95% of the requests from the project management group to add more fields and data to the warehouse were accommodated and the rest were in the process of being incorporated. Any inflexibility to accommodate new business needs and changes was due to resource constraints.

At Raymond James Financial, certain areas of the data warehouse responded well to changes in business need and other areas did not. In areas where the data warehouse responded fairly quickly to a change in business need, once the need was communicated to the data warehouse team, the development team made the changes relatively fast. When there was a change in a business event, the data warehouse team made the event data available to the data warehouse and accessible to the users. On the other hand, a majority of the users viewed the data warehouse as being unresponsive to accommodate their changing business needs. They felt that the response of the data warehouse was slow. A response could take anywhere from 2 weeks to 7–8 months, depending on the status of the requests on the queue.

At Raymond James Financial not everyone views the flexibility of the data warehouse to changing business needs in the same way. There seems to be a disparity between the perception of senior management and data warehouse managers and the users' perspective. The senior management views the data warehouse as being responsive to changing business needs, as do the data warehouse managers. But the users are dissatisfied with the data warehouse's responsiveness to changing business needs and view the data warehouse as being generally unresponsive.

It appears that although the existing architecture of the data warehouse is flexible to changes, and the database, application middleware, and front-end tools have high scalability, the gap between emerging business needs and the data warehouse's ability to support them is debilitating and of concern for its users. The infrastructure lags are not purely technical in nature. Organizational issues like resource constraints in financial investments and human resources for the data warehouse project are preventing the changes in business needs from being accommodated in the data warehouse in a quick and satisfactory manner.

The issue of flexibility in planning is pertinent to the data warehouse environment. As business needs change over time, a successful data warehouse needs to be flexible enough to be responsive to it. Although the data warehouse at Raymond James Financial is flexible in architecture, it is not perceived as flexible by the users, based on the response times to their requests. The resulting rigidity, due to resource constraints, could frustrate the strategic initiative of the data warehouse.

Conclusion

This chapter presented a case study at Raymond James Financial. The case study thoroughly assesses the cluster of factors leading to strategic alignment of the data warehouse to business plans. This chapter also analyzed the case study to determine whether an alignment of the data warehouse to business strategy and plans had an impact on its successful adoption. The case study indicates that the factors that facilitate strategic alignment of the data warehouse also influence the successful adoption of the data warehouse. Joint responsibility between business and data warehouse managers, alignment between business plan and data warehouse plan, flexibility in the data warehouse framework, business user satisfaction, and technical integration of the data warehouse all have a positive influence on the strategic alignment and successful adoption of the data warehouse.

Chapter 11

How to Assess Strategic Alignment of a Data Warehouse

Once an organization has implemented a data warehouse, how does it analyze and implement strategic alignment of its data warehouse to its business strategies and goals? Assessment of a gap between the effectiveness of the data warehouse and its expected results is one way of evaluating the alignment of the data warehouse. This chapter presents an interview instrument that can be used to conduct a gap analysis of the strategic alignment of a data warehouse at any typical organization.

The strategic alignment gap analysis approach consists of three steps:

1. Perform the interviews to assess the gap in strategic alignment of the data warehouse.
2. Analyze the results from the interview instruments and determine the root cause of the misalignments. Analysis is accomplished through direct interpretation of the individual instances and through aggregation of instances to find patterns, seek significance, and arrive at generalizations.
3. Implement a comprehensive alignment plan to take necessary actions to improve the alignment of the data warehouse to organizational goals.

The approach taken in this interview instrument is in depth and comprehensive. It helps understand strategic alignment from the participants' point of view. The questions in the interview instruments explore each of the underlying questions discussed in this book, providing a pathway to study the gap between the data warehouse implementation and the business strategies and goals. This interview instrument and questionnaire is generic and applicable to most industries. It may be modified to suit the particularities of a specific industry.

The design of the questionnaire is based on the factors discussed in this book to align data warehouses to the business strategies and goals.

Design of the Interview Instruments

The following five factors, posed as questions, provide the background to the design of the interview questions:

Q-1: In practice, how is joint responsibility established between business and data warehouse managers?

Q-2: How are the data warehouse plans aligned to business strategy and business plans?

Q-3: What is the impact of flexibility in the data warehouse framework on its strategic alignment and success?

Q-4: How does business user involvement and satisfaction affect data warehouse strategic alignment and success?

Q-5: What is the impact of degree of technical integration of the data warehouse on its strategic alignment and success?

The questions presented in this chapter are divided into three groups to address each group of stakeholders in the data warehousing process: the senior business managers, the data warehouse managers or team, and the business users, respectively. In addition to these open-ended questions, each participant may fill out a five-point graded questionnaire for each of these questions. This serves to capture the degree of agreement or disagreement with the assessed factors and the degree of satisfaction or dissatisfaction of the interview participants.

Design of Interview Instruments for Senior Business Managers

At the outset of the interview, questions should be asked of the senior business manager to gather an overview of the organization, such as

■ Name, locations, and subsidiaries of the organization
■ History
■ Industry/business function
■ Employee strength

- Gross turnover/net income
- Organizational structure

Questions for Senior Business Managers

1. Over the last 5 years what are the major changes that have taken place in your business plans?
2. Does the data warehouse provide you with information you need and has that resulted in changing business direction?
3. How well does the data warehouse support (a) your business plans and (b) your organization's plans?
4. Do the data warehouse managers help you understand the advantages and the limitations of (a) what the data warehouse can do and (b) what they can currently do for your needs?
5. How good is the response of the data warehouse team to your needs?
6. How has the data warehouse responded to your changing needs?
7. How involved are you in the data warehouse investment decisions?
8. In your opinion, how involved are you and your colleagues (including CEO) in data warehouse decisions?
9. How often do data warehouse managers participate in your strategy meetings?
10. Are you aware of a team that is integrating business needs and strategy with the data warehouse? If yes, how effective is this cooperation?
11. What are the formal communication channels between you and data warehouse managers?
12. Given a choice, what expertise would you like to add to the data warehouse team?
13. Finally, in your opinion, has the data warehouse been successful? What factors do you think are responsible for the data warehouse success?

Questions for Senior Business Managers Regrouped According to Main Questions

Interview questions for the senior business managers, the data warehouse managers, and the business users are regrouped according to the main question they address to facilitate analysis of the data gathered during the interviews.

Q-1: In practice, how is joint responsibility established between business and data warehouse managers?
How involved are you in the data warehouse investment decisions?
In your opinion, how involved are you and your colleagues (including CEO) in data warehouse decisions?
How often do data warehouse managers participate in your strategy meetings?

Q-2: How are the data warehouse plans aligned to business strategy and business plans?

How well does the data warehouse support (a) your business plans and (b) your organization's plans?

Do the data warehouse managers help you understand the advantages and the limitations of (a) what the data warehouse can do and (b) what they can currently do for your needs?

Are you aware of a team that is integrating business needs and strategy with the data warehouse? If yes, how effective is this cooperation?

What are the formal communication channels between you and data warehouse managers?

Q-3: What is the impact of flexibility in the data warehouse framework on its strategic alignment and success?

Over the last 5 years what are the major changes that have taken place in your business plans?

Does the data warehouse provide you with information you need and has that resulted in changing business direction?

Q-4: How does business user involvement and satisfaction affect data warehouse strategic alignment and success?

How good is the response of the data warehouse team to your needs?

How has the data warehouse responded to your changing needs?

Q-5: What is the impact of degree of technical integration of the data warehouse on its strategic alignment and success?

Given a choice, what expertise would you like to add to the data warehouse team?

Finally, in your opinion, has the data warehouse been successful? What factors do you think are responsible for the data warehouse success?

Design of Interview Instruments for Data Warehouse Managers

At the outset of the interview, questions should be asked of the senior IT or data warehouse manager to gather an overview of the IT infrastructure and the data warehouse infrastructure. These questions might include

- How would you describe your company's data warehouse architecture?
- What are the applications supported by the data warehouse?
- What are the tools you are using in the data warehouse and why?
- How many users currently access the data warehouse?
- How do the users access the data warehouse (through what access tools)?

Table 11.1 Questionnaire to Be Completed by Senior Business Managers

	Strongly Agree	Agree	Neither Agree nor Disagree	Disagree	Strongly Disagree
1. The data warehouse strongly supports the business plans.					
2. The data warehouse drives business decisions.					
3. Senior management has a high level of commitment to the data warehouse project.					
4. Business managers are highly involved in the data warehouse investment decisions.					
5. Data warehouse managers are highly involved in corporate strategy.					
8. Cross-functional teams are highly active in the data warehouse project.					
9. There are established communication channels to facilitate better understanding.					
10. The data warehouse team is aware of the business plans and strategies.					
11. The data warehouse is responsive to a change in business needs.					
13. The data warehouse is successful.					

Questions for Data Warehouse Managers

1. How was the data warehouse architecture selected (enterprise-wide, data mart, other)?
2. Over the last 5 years what major changes have taken place in the data warehouse plans and strategies?
3. Over the last 5 years has the data warehouse architecture changed? If so, how?
4. Do users have to comply with the tools and outputs you give, or do you choose tools to get the output that users want?
5. What are your users' key requirements now?
6. How involved are you in corporate strategy decisions?
7. Over the last 5 years what are the major changes that have taken place in your organization's business plan?
8. How is the integration of business and data warehouse planning process achieved?
9. How is the data integrated from different systems across the organization?
10. How is the data warehouse architecture integrated into existing IT systems' architecture?
11. How is the performance of the data warehouse evaluated? (Does your evaluation primarily depend on feedback by users?)
12. In your opinion, how flexible is the data warehouse to accommodate new business needs or changes?
13. Do the database, application middleware, and front-end tools have high scalability?
14. Do the database, application middleware, and front-end tools have high availability?
15. What problems have you encountered in the data warehouse project?
16. Given an option, what are the things you would like to improve or change in the data warehouse? What need would these changes fulfill?
17. In your opinion, what factors are responsible for the data warehouse success?

Questions for Data Warehouse Managers Regrouped According to Main Questions

Q-1: In practice, how is joint responsibility established between business and data warehouse managers?

How involved are you in corporate strategy decisions?

Over the last 5 years what are the major changes that have taken place in your organization's business plan?

Q-2: How are the data warehouse plans aligned to business strategy and business plans?

How was the data warehouse architecture selected (enterprise-wide, data mart, other)?

Over the last 5 years what major changes have taken place in the data warehouse plans and strategies?

How is the integration of business and data warehouse planning process achieved?

Q-3: What is the impact of flexibility in the data warehouse framework on its strategic alignment and success?

Over the last 5 years has the data warehouse architecture changed? If so, how?

In your opinion, how flexible is the data warehouse to accommodate new business needs or changes?

Do the database, application middleware, and front-end tools have high scalability?

Q-4: How does business user involvement and satisfaction affect data warehouse strategic alignment and success?

Do the database, application middleware, and front-end tools have high availability?

Do users have to comply with the tools and outputs you give, or do you choose tools to get the output that users want?

What are your users' key requirements now?

Q-5: What is the impact of degree of technical integration of the data warehouse on its strategic alignment and success?

How is the data integrated from different systems across the organization?

How is the data warehouse architecture integrated into existing IT systems' architecture?

How is the performance of the data warehouse evaluated? (Does your evaluation primarily depend on feedback by users?)

What problems have you encountered in the data warehouse project?

Overall questions:

Given an option, what are the things you would like to improve or change in the data warehouse? What need would these changes fulfill?

In your opinion, what factors are responsible for the data warehouse success?

Table 11.2 Questionnaire to Be Completed by Data Warehouse Managers

	Strongly Agree	Agree	Neither Agree nor Disagree	Disagree	Strongly Disagree
1. Data warehouse managers are aware of the corporate strategies.					
2. Data warehouse managers are highly involved in corporate strategy.					
3. Data warehouse plans support the business plans and strategies.					
4. Business decisions are the driver for the data warehouse design.					
5. Business and data warehouse planning processes are integrated.					
6. Business visions are the drivers for data warehouse architecture.					
7. Data is integrated from different systems across the organization.					
8. The data warehouse architecture is integrated into existing IT systems' architecture.					
9. The data warehouse technology was evaluated after the decision to build it.					
10. The data warehouse is highly responsive to a change in business needs.					
11. The database, application middleware, and front-end tools have scalability.					
12. The database, application middleware, and front-end tools have high availability.					
13. The data warehouse is successful.					

Design of Interview Instruments for Business Users

Questions for Business Users

1. Were you involved in the data warehouse project, and how did you participate?
2. How well does the data warehouse support your needs?
3. How are your needs communicated to the data warehouse team and vice versa? What are the formal communication channels between you and the data warehouse team?
4. How does the data warehouse respond to a change in business needs?
5. Does the data warehouse provide accurate information?
6. Does the data warehouse provide consistent and reliable information?
7. Does the data warehouse provide timely information?
8. Is the data warehouse easy to use?
9. Does the data warehouse enable day-to-day decisions?
10. Are the data warehouse functions and technical features easy to understand?
11. Was the user training adequate to carry out your functions?
12. Is the data warehouse project team highly skilled to manage and solve technical problems? How good is the response of the data warehouse team to your needs?
13. In your opinion, what factors are responsible for the data warehouse success?

Questions for Business Users Regrouped According to Main Questions

Q-1: In practice, how is joint responsibility established between business and data warehouse managers?

Q-2: How are the data warehouse plans aligned to business strategy and business plans?
   How are your needs communicated to the data warehouse team and vice versa? What are the formal communication channels between you and the data warehouse team?

Q-3: What is the impact of flexibility in the data warehouse framework on its strategic alignment and success?
   How does the data warehouse respond to a change in business need?

Q-4: How does business user involvement and satisfaction affect data warehouse strategic alignment and success?
   Were you involved in the data warehouse project, and how did you participate?
   How well does the data warehouse support your needs?
   Is the data warehouse easy to use?

Does the data warehouse enable day-to-day decisions?

Are the data warehouse functions and technical features easy to understand?

Was the user training adequate to carry out your functions?

Q-5: What is the impact of degree of technical integration of the data warehouse on its strategic alignment and success?

Does the data warehouse provide accurate information?

Does the data warehouse provide consistent and reliable information?

Does the data warehouse provide timely information?

Is the data warehouse project team highly skilled to manage and solve technical problems? How good is the response of the data warehouse team to your needs?

Overall question:

In your opinion, what factors are responsible for the data warehouse success?

Table 11.3   Questionnaire to Be Completed by Business Users

	Highly Satisfied	Satisfied	Moderately Satisfied	Dissatisfied	Highly Dissatisfied
1. Users' participation in the data warehouse project					
2. Communication of users' needs to the data warehouse team					
3. Communication by the data warehouse team to the users					
4. Data warehouse response to change in business needs					
5. Accuracy of data warehouse information					
6. Consistency and reliability of the data warehouse information					
7. Timeliness of data warehouse information					
8. Ease of use of the data warehouse					
9. Relevance of the data warehouse information to day-to-day decisions					
10. Users' understanding of the data warehouse functions and features					
11. Adequacy of user training					
12. The data warehouse project team's skill to manage and solve technical problems, response of the data warehouse team					
13. Level of satisfaction with the data warehouse success					

Chapter 12

Epilogue

This book presents a pathway for strategic alignment of the data warehouse. The methods discussed in the book are designed to aid data warehouse planners to align the data warehouse to organizational goals and objectives. To harness the power of data warehousing for long-term benefits, it is important, during the entire process, to continuously recognize the dimensions, interplay, and dependence of (a) strategy, (b) organizational culture, (c) service orientation, and (d) the governance process inherent to the success of this effort.

Strategy

With enhanced need in organizations for integration of new and existing systems, as well as due to resource constraints, data warehousing is assuming an increasingly strategic role. Yet a large number of data warehouse initiatives either fail or do not meet the expected level of outcomes (Figure 12.1).

Several possible reasons for this performance gap have been discussed in this book. Defects in both strategic planning and strategic execution are often responsible for failure of data warehouse strategies (Figure 12.2).

Lack of understanding and poor choices by senior management result in pressing for better execution when there is a need for a better strategy. At other times senior management is insisting on a change in direction when the need is for greater focus on execution. Several steps can be taken for better strategic alignment of the data warehouse. Tracking the data warehouse performance against long-term plans would enable better planning and execution. Putting in place a governance and feedback system would allow for adjustments to changes in organizational goals.

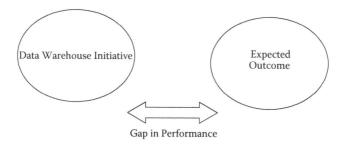

Figure 12.1   Data warehouse performance gap.

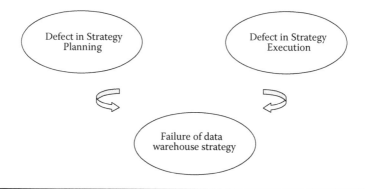

Figure 12.2   Failure of data warehouse strategy.

Often strategies linking the data warehouse to organizational goals are approved by senior management but are not well communicated within the organization. Ineffective communication makes it difficult to translate the strategy to actionable plans and allocate resources to execute the plans (Figure 12.3).

The operational levels within the organization remain unaware of what is required to meet the expectations of senior management. Better communication and linking resource allocation to strategy planning and development would ensure attainment of data warehouse goals. Well-formulated strategic plans for the data warehouse, appropriate resource application, good communication channels, and specified accountability for results would all facilitate better strategic alignment, resource and action planning, and successful implementation of data warehouses (Figures 12.4 and Figure 12.5).

Organizational Culture

Organizational culture has a significant influence on the effectiveness of the data warehouse implementation and strategic alignment. Organizational culture is

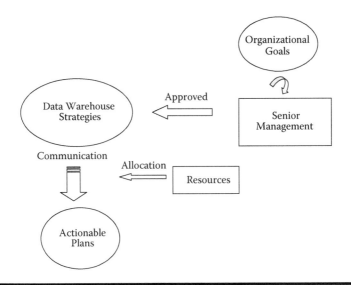

Figure 12.3   Translation of data warehouse strategy.

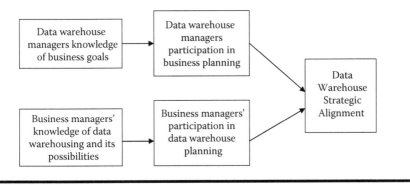

Figure 12.4   Data warehouse strategic alignment.

reflected in the core values, beliefs, and assumptions as well as the norms and practices of the organization. Data warehouses store and share information within the organization. Data warehousing activities lead to enhanced communications and an increase in the level of participation across the organization's functional boundaries. Values and work systems that exist in the organization can either encourage or restrain the activities and related outcomes of data warehousing (Figure 12.6).

The data warehouse is not an independent phenomenon in the organization. Its success among the users depends on the credibility and likeability of the data warehouse leadership. The data warehouse in the organization is embedded in a social context that determines how the owners of the data share the information as well as how the users of the data warehouse interpret the information. Organizational beliefs

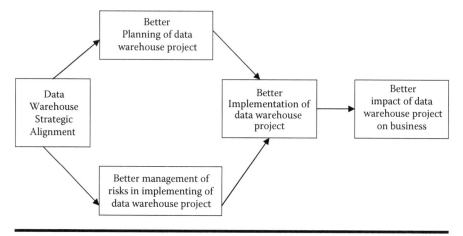

Figure 12.5 Impact of data warehouse strategic alignment.

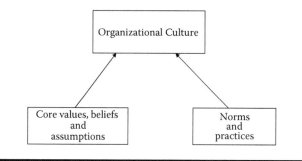

Figure 12.6 Organizational culture.

are formed over time with shared experiences of coping with problems and devising solutions. Groups within an organization may perceive situations and activities differently, influenced by their shared values. These values often define the context of interaction within the different groups in the organization. Different groups within the organization perceive technology differently, and these embedded values lead to different patterns of use of data warehouses. The data warehouse managers cannot expect uniformity in how the users respond to the data warehouse.

Even though there may be different local group cultures within an organization, a dominant unifying organizational culture can motivate people toward a common action. A collaborative environment within the organization can influence the behaviors of different functional groups and facilitate knowledge transfer among the various stakeholders of the data warehouse. A collaborative atmosphere also leads to efficiencies in problem solving and improvements in teamwork, leading

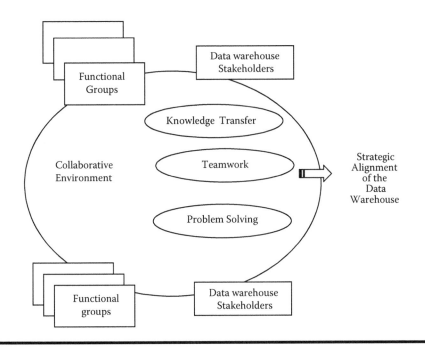

Figure 12.7   Collaborative organizational culture.

to better performance of data warehousing efforts. An environment where the different functional departments compete and conflict with each other can make for a potentially difficult implementation and strategic alignment of a data warehouse (Figure 12.7).

Organizational culture plays an important role in managing, using, and aligning data warehouses effectively. Organizations with an open and supportive culture predispose people to more constructive behaviors. Data warehouses functioning in such environments have the ability to respond more rapidly to changes in business strategies and user demands. A trusting and collaborative approach leads to a greater willingness among the data warehouse stakeholders to cooperate and share expertise. Organizational cultures that thrive on power tend to display more knowledge-hoarding behaviors and consequently less shared information and data. In such situations, senior management may need to set cultural change efforts in motion. Senior leadership could help initiate programs to develop relationships between departments and encourage participation in organization-wide social events to develop connections across functional boundaries. Data warehouse leadership is essential to the success of the data warehouse. Senior leadership should encourage data warehouse leaders and data warehouse stakeholders who exhibit expertise and collaborative spirit.

Service Orientation of Data Warehouses

Building data warehouses requires a close interaction between the customer of the data warehouse (the business user) and the provider of the data warehouse (the data warehouse team). It requires the exchange and combination of knowledge to build a useful system. The provider of the data warehouse may lack the subject matter expertise and contextual knowledge of the business user and how the business user is going to leverage the data warehouse to compete effectively. The business user, on the other hand, may lack the knowledge of the full capabilities that the data warehouse technology can provide and the experience of the data warehouse team from other similar projects on what will work best. Although information asymmetry exists as part of any service relationship, the intangibility of the service and the scale of information in a data warehouse introduce new levels of complexity in the coordination and development of data warehouses.

Building data warehouses involves the combining of both codified and tacit knowledge (Figure 12.8). The tacit knowledge involved in the building of data warehouses limits the ability of both the business user and the data warehousing team to fully comprehend the needs and capability of the other. During the building of enterprise-wide data warehouses, the profusion of information and the necessity to integrate tacit knowledge into rational solutions presents a big challenge. The data warehouse provides the opportunity to configure enterprise-wide information to create new value. The tasks of coordinating resources, integrating information, and combining, customizing, and reusing information is complicated and tremendous. It involves systems integration, business process remodeling, and organizational change.

The complexity described here can be managed when both parties interact closely in the design and building of the data warehouse. When the business user is a co-producer of the data warehouse and is involved in defining, designing, and integrating the information, the co-generation of the data warehouse leads to much greater chances of success and alignment of the data warehouse.

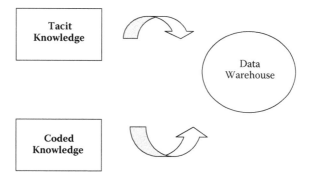

Figure 12.8 Knowledge integration.

Governance Process

The making of strategic decisions regarding data warehousing in organizations can be planned or ad hoc (often due to changes in leadership). These decisions are made by stakeholders, persons of influence in the organization, and those involved in the planning and implementation of the data warehouse. A governance process ensures a smoother and more successful implementation of the data warehouse and its alignment to business goals and strategies (Figure 12.9). The governance process ensures that responsibility and accountability is delegated for the data warehouse process. Carefully structured governance processes allow the achievement of a coordinated and effective data warehouse strategy implementation. Governance ensures the devising of appropriate processes, structures, and policies to facilitate effective decision making and implementation.

Strategic decision making for the data warehouse should encompass the needs of a wide spectrum of interested parties. Differing needs of data warehouse stakeholders must be acknowledged and planning must be flexible to reconcile conflicting interests. When data warehouse decisions are made and endorsed by the stakeholders, there is greater buy-in, and this ensures actions on the decisions reached. Responsiveness to data warehouse stakeholders' interests enhances the ability to recognize and manage a greater range of risks. The incorporation of a feedback mechanism in the governance process makes way for continuous improvement and realignment of the data warehouse strategy with organizational objectives.

Data warehouse governance can meet the organizational goals through the development and effective implementation of the data warehouse, alignment of the data warehouse with the organizational strategy, and regular review, evaluation, and monitoring of the data warehouse infrastructure and human capital (Figure 12.10). Governance of data warehouses covers the realm of infrastructure, architecture, and prioritization strategies. Governance involves formulation of strategy, risk management, delivering financial value, and performance measurement (Figure 12.11). The formulation of data warehouse strategy in an organization

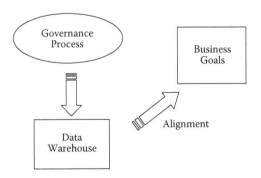

Figure 12.9 Data warehouse governance and alignment.

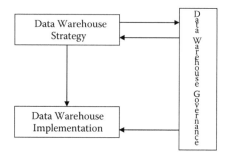

Figure 12.10   Data warehouse governance of DW strategy and implementation.

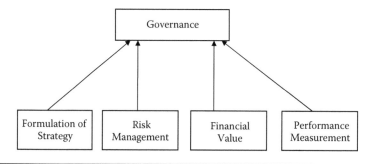

Figure 12.11   Governance process.

meets the need to integrate and disseminate information to meet organizational objectives and goals while managing risks. The governance process ensures that the data warehouse delivers value to its stakeholders. The data warehouse strategy cannot be static. It has to be flexible and responsive to adjustments made in organizational strategies. The data warehouse strategy cannot be a formula that fits all organizations or all occasions.

Effective data warehouse governance, both long and short term, can leverage organizational data and information into a strategic asset and a competitive differentiator. Data warehouse governance can support an effective understanding of the role and potential of the data warehouse within the organization.

With the advent of globalization, outsourcing parts or the whole of data warehouse projects is a distinct possibility. Outsourcing not only creates value but also involves risks. Governance plays an important role in managing these risks in data warehousing projects. Governance manages the interdependence among tasks when a particular function is outsourced. Sound governance allows coordinations and joint problem solving between the small retained data warehouse team within the organization and the outsourcing company.

References

Agosta, L. (1999), *The Essential Guide to Data Warehousing*. Upper Saddle River, NJ: Prentice Hall.

Agrawal, D., A. El Abbadi, A. Singh, and T. Yurek (1997), "Efficient view maintenance at data warehouses," *ACM SIGMOD International Conference on Management of Data*, Tucson, AZ, 26 (2), 417–427.

Alavi, M. and D.E. Leidner (1999), "Knowledge management systems: Issues, challenges, and benefits," *Communications of the AIS*, 1 (2es) 7, 1–37.

Altinkemer, K. (2001), "Bundling E-banking services," *Communications of the ACM,* 44(6), 45–47.

Anahory, S. and D. Murray (1997), *Data Warehousing in the Real World: A Practical Guide for Building Decision Support Systems*. Boston: Addison-Wesley Longman Publishing.

Andal-Ancion, A., P.A. Cartwright, and G.S. Yip (2003), "The digital transformation of traditional businesses," *Sloan Management Review*, 44 (4), 34–41.

Ang, J. and T.S.H. Teo (2000), "Management issues in data warehousing: Insights from the Housing and Development Board," *Decision Support Systems*, 29 (1), 11–20.

Anton, J. (2000), "The past, present and future of customer access centers," *International Journal of Service Industry Management*, 11 (2), 120–130.

Armstrong, C.P. and V. Sambamurthy (1999), "Information technology assimilation in firms: The influence of senior leadership and IT infrastructures," *Information Systems Research*, 10 (4), 304–327.

Armstrong, R. (1997), "Data warehousing: Dealing with the growing pains," *Proceedings of the Thirteenth International Conference on Data Engineering*, Birmingham UK, 199–205.

Bai, R.J. and G.G. Lee (2003), "Organizational factors influencing the quality of the IS/IT strategic planning process," *Industrial Management and Data Systems*, 103 (8), 622–632.

Bakos, J.Y. and M.E. Treacy (1986), "Information technology and corporate strategy: A research perspective," *MIS Quarterly*, 10 (2), 107–119.

Ballou, D.P. and G.K. Tayi (1999), "Enhancing data quality in data warehouse environments," *Communications of the ACM*, 42 (1), 73–78.

Beeson, I. and S. Al Mahamid (2003), "Survey of strategic alignment indicators in manufacturing companies in the South-West of England," http://www.cems.uwe.ac.uk/~xzhang/CRES2003/papers/SoudM.pdf.

Benander, A., B. Benander, A. Fadlalla, and G. James (2000), "Data warehouse administration and management," *Information Systems Management*, 17 (1), 71–80.

Bernardino, J. and H. Madeira (2000), "Data warehousing and OLAP: Improving query performance using distributed computing," *CAiSE*00 Conference on Advanced Information Systems Engineering*, Sweden.

Berndt, D.J., A.R. Hevner, and J. Studnicki (2003), "The CATCH data warehouse: Support for community health care decision-making," *Decision Support Systems*, 35 (3), 367–384.

Berndt, D.J. and R.K. Satterfield (2000), "Customer and household matching: Resolving entity identity in data warehouses," *Proceedings of SPIE — The International Society for Optical Engineering*, 4057, 173–180.

Berry, M. and G. Linoff (1999), *Mastering Data Mining: The Art and Science of Customer Relationship Management*. New York: John Wiley.

Berry, M.J.A. and G.S. Linoff (2004), *Data Mining Techniques: For Marketing, Sales, and Customer Relationship Management*. New York: John Wiley.

Berson, A., S. Smith, and K. Thearling (2000), *Building Data Mining Applications for CRM*. New York: McGraw-Hill.

Bharadwaj, A.S. (2000), "A resource-based perspective on information technology capability and firm performance: An empirical investigation," *MIS Quarterly*, 24 (1), 169–196.

Bhowmick, S.S., S. Madria, and W.K. Ng (2003a), "Representation of Web data in a Web warehouse," *The Computer Journal*, 46 (3), 229–262.

Bhowmick, S.S., S.K. Madria, and W.K. Ng (2004), *Web Data Management: A Warehouse Approach*. Cambridge, MA: Springer.

Bhowmick, S.S., W.K. Ng, and S. Madria (2003b), "Constraint-driven join processing in a Web warehouse," *Data and Knowledge Engineering*, 45 (1), 33–78.

Bliujute, R., S. Saltenis, G. Slivinskas, and C.S. Jensen (1998), "Systematic change management in dimensional data warehousing," *Proceedings of the Third International Baltic Workshop on DB and IS*, Riga, Latvia, 27–41.

Boddy, D. and R. Paton (2005), "Maintaining alignment over the long-term: lessons from the evolution of an electronic point of sale system," *Journal of Information Technology*, 20(3), 141.

Bourgeois, L.J. and D.R. Brodwin (1984), "Strategy implementation: Five approaches to an elusive phenomenon," *Strategic Management Journal*, 5, 241–264.

Boynton, A.C. and R.W. Zmud (1987), "Information technology planning in the 1990's: Directions for practice and research," *MIS Quarterly*, 11 (1), 59–71.

Brazelton, J. and G.A. Gorry (2003), "On site: Creating a knowledge-sharing community: If you build it, will they come?," *Communications of the ACM*, 46 (2), 23–25.

Breathnach, P. (2000), "Globalisation, information technology and the emergence of niche transnational cities: The growth of the call centre sector in Dublin," *Geoforum*, 31 (4), 477–485.

Bresnahan, T.F., E. Brynjolfsson, and L.M. Hitt (2002), "Information technology, workplace organization, and the demand for skilled labor: Firm-level evidence," *The Quarterly Journal of Economics*, 117 (1), 339–376.

Bruce, K. (1998), "Can you align IT with business strategy," *Strategy and Leadership*, 26 (5), 16–21.

Brynjolfsson, E. and L.M. Hitt (1995), "Information technology as a factor of production: The role of differences among firms," *Economics of Innovation and New Technology*, 3–4, 183–199.

Brynjolfsson, E. and L.M. Hitt (2000), "Beyond computation: Information technology, organizational transformation and business performance," *The Journal of Economic Perspectives*, 14 (4), 23–48.

Burn, J. M. (1996), "IS innovation and organizational alignment — a professional juggling act," *Journal of Information Technology*, 11 (1), 3–12.

Burn, J.M. and C. Szeto (2000), "A comparison of the views of business and IT management on success factors for strategic alignment," *Information and Management*, 37 (4), 197–216.

Bussmann, S. (1998), "An agent-oriented architecture for holonic manufacturing control," *Proceedings of First International Workshop on IMS*, Lausanne, Switzerland, 1–12.

Buzydlowski, J.W., I.Y. Song, and L. Hassell (1998), "A framework for object-oriented on-line analytic processing," *Proceedings of the 1st ACM International Workshop on Data Warehousing and OLAP*, Washington, DC, 10–15.

Calantone, R.J. and C.A. Di Benedetto (2000), "Performance and time to market: Accelerating cycle time with overlapping stages," *IEEE Transactions on Engineering Management*, 47 (2), 232–244.

Caldow, J.C. and J.B. Kirby (1996), "Business culture: The key to regaining competitive edge," in *Competing in the Information Age: Strategic Alignment in Practice*, J.N. Luftman (Ed.), 10, 293–321.

Calvanese, D., G. De Giacomo, M. Lenzerini, D. Nardi, and R. Rosati (1998), "Source integration in data warehousing," *Proceedings of Ninth International Workshop on Database and Expert Systems Applications*, Vienna, Austria, 192–197.

Calvanese, D., G. De Giacomo, M. Lenzerini, D. Nardi, and R. Rosati (1999), "A principled approach to data integration and reconciliation in data warehousing," *Proceedings of the International Workshop on Design and Management of Data Warehouses*, Heidelberg, Germany, http://sunsite.informatik.rwth-aachen.de/Publications/CEUR-WS/Vol-19/.

Carr, N.G. (2003), "IT Doesn't Matter," *Harvard Business Review*, 81 (5), 41–49.

Chaffey, D., R. Mayer, K. Johnston, and F. Ellis-Chadwick (2000), *Internet Marketing: Strategy, Implementation and Practice*. London: Prentice Hall.

Chan, Y.E. (1996), *Business Strategic Orientation, Information Systems Strategic Orientation, and Strategic Alignment*. Cambridge, MA: Marketing Science Institute.

Chan, Y.E. and S. Huff (1993a), "Strategic information systems alignment," *Business Quarterly*, 58 (1), 51–56.

Chan, Y.E. and S.L. Huff (1993b), "Investigating information systems strategic alignment," *Proceedings of the Fourteenth International Conference on Information Systems*, 345–363.

Chan, Y.E., S.L. Huff, D.W. Barclay, and D.G. Copeland (1997), "Business strategy orientation, information systems orientation and strategic alignment," *Information Systems Research*, 8 (2), 125–150.

Chau, K.W., Y. Cao, M. Anson, and J. Zhang (2003), "Application of data warehouse and decision support system in construction management," *Automation in Construction*, 12 (2), 213–224.

Chaudhuri, S. and U. Dayal (1997), "An overview of data warehousing and OLAP technology," *ACM SIGMOD Record*, 26 (1), 65–74.

Chen, I.J. and K. Popovich (2003), "Understanding customer relationship management (CRM)," *Business Process Management Journal*, 9, 672–688.

Chen, L., K.S. Soliman, E. Mao, and M.N. Frolick (2000), "Measuring user satisfaction with data warehouses: An exploratory study," *Information and Management*, 37 (3), 103–110.

Chenoweth, T., K. Corral, and H. Demirkan (2006), "Seven key interventions for data warehouse success," *Communications of the ACM*, 49 (1), 114–119.

Cooper, B.L., H.J. Watson, B.H. Wixom, and D.L. Goodhue (2000), "Data warehousing supports corporate strategy at First American Corporation," *MIS Quarterly*, 24 (4), 547–567.

Counihan, A., P. Finnegan, and D. Sammon (2002), "Towards a framework for evaluating investments in data warehousing," *Information Systems Journal*, 12 (4), 321–338.

Cui, Y. and J. Widom (2003), "Lineage tracing for general data warehouse transformations," *The International Journal on Very Large Data Bases*, 12 (1), 41–58.

Date, C.J. (1999), *An Introduction to Database Systems*. Boston: Addison-Wesley Longman Publishing.

Datta, A., B. Moon, and H. Thomas (1998), "A case for parallelism in data warehousing and OLAP," *Ninth International Workshop on Database and Expert Systems Applications*, 98, 226–231.

Davenport, T.H. (1998), "Putting the enterprise into the enterprise system," *Harvard Business Review*, 76 (4), 121–131.

Davis, L., B. Dehning, and T. Stratopoulos (2003), "Does the market recognize IT-enabled competitive advantage?," *Information and Management*, 40 (7), 705–716.

Dean Jr., J.W. and M.P. Sharfman (1996), "Does decision process matter? A study of strategic decision-making effectiveness," *The Academy of Management Journal*, 39 (2), 368–396.

Debevoise, T. (1999), *The Data Warehouse Method: Integrated Data Warehouse Support Environments*. Upper Saddle River, NJ: Prentice Hall.

Dehne, F., T. Eavis, and A. Rau-Chaplin (2003), "Parallel multi-dimensional ROLAP indexing," *Proceedings of 3rd IEEE/ACM International Symposium on Cluster Computing and the Grid*, Tokyo, Japan, 86–93.

DeLone, W.H. and E.R. McLean (1992), "Information systems success: The quest for the dependent variable," *Information and Management*, 3, 60–95.

Dembo, R. (2004), "The risk architect: Integrating risk and finance," *Risk*, 17 (11), 74.

Dinter, B., C. Sapia, G. Höfling, and M. Blaschka (1998), "The OLAP market: State of the art and research issues," *1st ACM International Workshop on Data Warehousing and OLAP*, Washington, DC, 22–27.

Doherty, N.F. and G. Doig (2003), "An analysis of the anticipated cultural impacts of the implementation of data warehouses," *IEEE Transactions of Engineering Management*, 50 (1).

Dyche, J. (2002), *The CRM Handbook: A Business Guide to Customer Relationship Management*. Reading, MA: Addison-Wesley Professional.

Edwards, B.A. (2000), "Chief executive officer behavior: The catalyst for strategic alignment," *International Journal of Value-Based Management*, 13 (1), 47–54.

Eisenhardt, K.M. and C.B. Schoonhoven (1996), "Resource-based view of strategic alliance formation: Strategic and social effects in entrepreneurial firms," *Organization Science*, 7 (2), 136–150.

English, L.P. (1999), *Improving Data Warehouse and Business Information Quality: Methods for Reducing Costs and Increasing Profits*. New York: John Wiley.

Ester, M., H.P. Kriegel, J. Sander, M. Wimmer, and X. Xu (1998), "Incremental clustering for mining in a data warehousing environment," *Proceedings of the 24th International Conference on Very Large Databases*, New York.

Eustace, C. (2003), "A new perspective on the knowledge value chain," *Journal of Intellectual Capital*, 4 (4), 588–596.

Fiedler, K.D., V. Grover, and J.T.C. Teng (1994), "Information technology-enabled change: The risks and rewards of business process redesign and automation," *Journal of Information Technology*, 9 (4), 267–275.

Fisher, C.W., I.S. Chengalur-Smith, and D.P. Ballou (2003), "The impact of experience and time on the use of data quality information in decision making," *Information Systems Research*, 14 (2), 170–188.

Fisher, M. and A. Raman (1996), "Reducing the cost of demand uncertainty through accurate response to early sales," *Operations Research*, 44 (1), 87–99.

Freude, R. and A. Konigs (2003), "Tool integration with consistency relations and their visualisation," *Proceedings of ESEC/FSE Workshop on Tool-Integration in System Development*, Helsinki, 6–10.

Frolick, M.N. and K. Lindsey (2003), "Critical factors for data warehouse failure," *Journal of Data Warehousing*, 8 (1), 48–54.

Gagnon, S. (1999), "Resource-based competition and the new operations strategy," *International Journal of Operations and Production Management*, 19 (2), 125–138.

Galbraith, J.R. and R.K. Kazanjian (1986), *Strategy Implementation: Structure, Systems, and Process*. St. Paul, MN: West Publishing.

Gardner, S.R. (1998), "Building the data warehouse," *Communications of the ACM*, 41 (9), 52–60.

Gary, C. (2004), "The evolving enterprise data warehouse market, part 1," *META DELTA*, Vol. Delta 2776.

Gatziu, S., M. Jeusfeld, M. Staudt, and Y. Vassiliou (1999), "Design and management of data warehouses. Report on the DMDW'99 workshop," Heidelberg, Germany, 7–10.

Genesereth, M.R., A.M. Keller, and O.M. Duschka (1997), "Infomaster: An information integration system," *Proceedings of the International Workshop "Intelligent Information Integration" during the 21st German Annual Conference on Artificial Intelligence*, Freiburg, Germany, September 1997.

Glazer, R. (1991), "Marketing in an information-intensive environment: Strategic implications of knowledge as an asset," *Journal of Marketing*, 55 (4), 1–19.

Glazer, R. (1993), "Measuring the value of information: The information-intensive organization," *IBM Systems* Journal, 32 (1), 99–110.

Goldstein, H. (2005), "Who killed the virtual case file?," *IEEE Spectrum*, 42 (8).

Golfarelli, M. and S. Rizzi (1998), "A methodological framework for data warehouse design," *Proceedings of the 1st ACM International Workshop on Data Warehousing and OLAP*. Washington, DC, 3–9.

Gorla, N. (2003), "Features to consider in a data warehousing system," *Communications of the ACM*, 46 (11), 111–115.

Gray, P. and H.J. Watson (1998), "Present and future directions in data warehousing," *The Database for Advances in Information Systems*, 29 (3), 83.

Griffin, A. (1997), "The effect of project and process characteristics on product development cycle time," *Journal of Marketing Research*, 34 (1), 24–35.

Groth, R. (2000), *Data Mining: Building Competitive Advantage*. Upper Saddle River, NJ: Prentice Hall PTR.

Guimaraes, T., D.S. Staples, and J.D. McKeen (2003), "Empirically testing some main user-related factors for systems development quality," *The Quality Management Journal*, 10 (4), 39–54.

Gupta, A.K. and V. Govindarajan (1984), "Business unit strategy, managerial character-istics, and business unit effectiveness at strategy implementation," *The Academy of Management Journal*, 27 (1), 25–41.

Gupta, A., V. Harinarayan, and D. Quass (1995), "Aggregate-query processing in data ware-housing environments," *21st International Conference on Very Large Databases*, Zurich.

Gupta, A. and I.S. Mumick (1995), "Maintenance of materialized views: Problems, tech-niques, and applications," *Data Engineering Bulletin*, 18 (2), 3–18.

Gupta, H. and I.S. Mumick (2005), "Selection of views to materialize in a data warehouse," *IEEE Transactions on Knowledge and Data Engineering*, 17 (1), 24–43.

Hammer, J., H. Garcia-Molina, J. Widom, W. Labio, and Y. Zhuge (1995), "The Stanford Data Warehousing Project," *Data Engineering Bulletin*, 18 (2), 41–48.

Han, J., N. Stefanovic, and K. Koperski (1998), "Selective materialization: An efficient method for spatial data cube construction," *Conference on Knowledge Discovery and Data Mining (PAKDD'98)*. Melbourne, Australia.

Henderson, J.C. and N. Venkatraman (1989), *Strategic Alignment: A Process Model for Integrating Information Technology and Business Strategies*. Cambridge, MA: Massachu-setts Institute of Technology, Sloan School of Management, Center for Information Systems Research.

Henderson, J.C. and N. Venkatraman (1990a), "Strategic alignment: A framework for research on the strategic management of information technology," *MIT Behavioral and Political Sciences Conference on Organizational Change*.

Henderson, J.C. and N. Venkatraman (1990b), *Strategic Alignment: A Model for Organi-zational Transformation via Information Technology*. Cambridge, MA: Massachusetts Institute of Technology, Sloan School of Management, Center for Information Systems Research.

Henderson, J.C., N. Venkatraman, and S. Oldach (1996), "Aligning business and IT strate-gies," in *Competing in the Information Age: Strategic Alignment in Practice*, Jerry N. Luftman (Ed.). New York: Oxford University Press.

Hirschheim, R. and R. Sabherwal (2001), "Detours in the path toward strategic information systems alignment," *California Management Review*, 44 (1), 87–108.

Hitt, J.C. (2001), "Connecting IT possibilities and institutional priorities," *Educase Review*, 36 (6), 8–9.

Ho, D.C.K., K.F. Au, and E. Newton (2003), "The process and consequences of supply chain virtualization," *Industrial Management and Data Systems*, 103 (6), 423–433.

Hoven, J. van den (1008), "Data marts: Plan big, build small," *Information Systems Management*. 15:1, 1–3.

Hristovski, D., M. Rogac, and M. Markota (2000), "Using data warehousing and OLAP in public health care," *Proceedings from AMIA Symposium*, Slovenia, Vol. 369.

Hu, Q. and J. Quan (2003), "Information intensity and impact of IT investments on pro-ductivity: An industry level perspective," *Proceedings of the 11th European Conference on Information Systems*. Naples, Italy.

Hui, S.C. and G. Jha (2000), "Data mining for customer service support," *Information and Management*, 38 (1), 1–13.

Hurd, M. (2003), "Enterprise decision making," *Baylor Business Review*, 21 (1), 10–13.

Husemann, B., J. Lechteneborger, and G. Vossen (2000), "Conceptual data warehouse design," *Proceedings of the International Workshop on Design and Management of Data Warehouses*, Stockholm, Sweden, June 5–6, 6-1-6-11.

Hwang, H.G., C.Y. Ku, D.C. Yen, and C.C. Cheng (2004), "Critical factors influencing the adoption of data warehouse technology: A study of the banking industry in Taiwan," *Decision Support Systems*, 37 (1), 1–21.

IBM (1999), "Meta-data management for Business Intelligence Solutions — IBM's solutions," in *Data Warehousing*, SCN Education B.V. (Ed.). Germany: Vieweg.

Ingham, J. (2000), "Data warehousing: A tool for the outcomes assessment process," *IEEE Transactions on Education*, 43 (2), 132–136.

Inmon, W.H. (1993), "The operational data store," *PRISM Tech Topic*, Vol. 1.

Inmon, W.H. (1996a), *Building the Data Warehouse*. New York: John Wiley.

Inmon, W.H. (1996b), "The data warehouse and data mining," *Communications of the ACM*, 39 (11), 49–50.

Inmon, W.H. (2005), *Building the Data Warehouse, 2nd Edition,* New York: John Wiley.

Ives, B. and G.P. Learmonth (1984), "The information system as a competitive weapon," *Communications of the ACM*, 27 (12), 1193–1201.

James, J. (1999), *Globalization, Information Technology and Development*. New York: St. Martin's Press.

Jarvenpaa, S.L. and B. Ives (1991), "Executive involvement and participation in the management of information technology," *MIS Quarterly*, 15 (2), 205–227.

Jiang, J.J., W.A. Muhanna, and G. Klein (2000), "User resistance and strategies for promoting acceptance across system types," *Information and Management*, 37 (1), 25–36.

Johnson, G. and K. Scholes (1993), *Exploring Corporate Strategy*. New York: Prentice Hall.

Johnson, L.K. (2004), "Strategies for data warehousing," *MIT Sloan Management Review* (Spring).

Jones, N. and T. Kochtanek (2002), "Consequences of Web-based technology usage," *Online Information Review*, 26 (4), 256–264.

Kandampully, J. and R. Duddy (1999), "Competitive advantage through anticipation, innovation and relationships," *Management Decision*, 37 (1), 51–56.

Kaplan, R.S. and D.P. Norton (1992), "The balanced scorecard: Measures that drive performance," *Harvard Business Review*, 70 (1) 71–79.

Kärkkäinen, M. and J. Holmström (2002), "Wireless product identification: Enabler for handling efficiency, customisation and information sharing," *Supply Chain Management: An International Journal*, 7 (4), 242–252.

Katic, N., G. Quirchmay, J. Schiefer, M. Stolba, and A.M. Tjoa (1998), "A prototype model for data warehouse security based on metadata," *Proceedings of the Ninth International Workshop on Database and Expert Systems Applications*. Vienna, Austria.

Kayworth, T. R., D. Chatterjee, and V. Sambamurthy (2001), "Theoretical justification for IT infrastructure investments," *Information Resources Management Journal*, 14 (3), 5–14.

Kearns, G.S. and A.L. Lederer (2000), "The effect of strategic alignment on the use of IS-based resources for competitive advantage," *Journal of Strategic Information Systems*, 9 (4), 265–293.

Kearns, G.S. and A.L. Lederer (2003), "A resource-based view of strategic IT alignment: How knowledge sharing creates competitive advantage," *Decision Sciences*, 34 (1), 1–29.

Kelly, S. (1997), *Data Warehousing in Action*. New York: John Wiley.

Kimball, R. (1996), *The Data Warehouse Toolkit: Practical Techniques for Building Dimensional Data Warehouses*. New York: John Wiley.

Kimball, R. (1998), *The Data Warehouse Lifecycle Toolkit* (1st ed.). New York: John Wiley.

Klenz, B. (2001), "Processing 'one version of truth' improves user confidence," *Control Engineering*, 48 (1), 47.

Koch, C. (1999), "The Middle Ground," *CIO Magazine*, Jan 15.

Kodama, M. (2002), "Strategic community management with customers: Case study on innovation using IT and multimedia technology in education, medical and welfare fields," *International Journal of Value-Based Management*, 15 (3), 203–224.

Kosala, R. and H. Blockeel (2000), "Web mining research: A survey," in *ACM SIGKDD Explorations Newsletter*. New York: ACM Press, 2.

Kotidis, Y. and N. Roussopoulos (1998), "An alternative storage organization for ROLAP aggregate views based on cubetrees," in *Proceedings of the 1998 ACM SIGMOD International Conference on Management of Data*. New York: ACM Press, 27.

Krishnan, M.S., V. Ramaswamy, M.C. Meyer, and P. Damien (1999), "Customer satisfaction for financial services: The role of products, services, and information technology," *Management Science*, 45 (9), 1194–1209.

Landry, Jr., R., R. Debreceny, and G.L. Gray (2004), "Grab your picks and shovels! There's gold in your data," *Strategic Finance*, 85 (7), 24–28.

Lederer, A.L. and A.L. Mendelow (1988), "Convincing top management of the strategic potential of information systems," *International Journal of Information Management*, 12 (4), 525–534.

Lee, G.G. and R.J. Bai (2003), "Organizational mechanisms for successful IS/IT strategic planning in the digital era," *Management Decision*, 41 (1), 32–42.

Lee, H., T. Kim, and J. Kim (2001), "A metadata oriented architecture for building data warehouse," *Journal of Database Management*, 12 (4), 15–25.

Lee, H.L. (2002), "Aligning supply chain strategies with product uncertainties," *California Management Review*, 44 (3), 105–119.

Lee, H.L. and S. Whang (2001), "E-business and supply chain integration," *Stanford Global Supply Chain Management Forum*, 2.

Lee, Y.W., L. Pipino, D.M. Strong, and R.Y. Wang (2004), "Process-embedded data integrity," *Journal of Database Management*, 15 (1), 87–103.

Li, C. and X.S. Wang (1996), "A data model for supporting on-line analytical processing," in *Proceedings of the Fifth International Conference on Information and Knowledge Management*. New York: ACM Press.

Little, R.G. and M.L. Gibson (2003), "Perceived influences on implementing data warehousing," *IEEE Transactions on Software Engineering*, 29 (4), 290–296.

Little, R.G.J. and M.L. Gibson (1999), "Identification of factors affecting the implementation of data warehousing," *Thirty-Second Annual Hawaii International Conference on System Sciences*, 7, 7011.

Loebbecke, C. and J. Wareham (2003), "The impact of ebusiness and the information society on 'STRATEGY' and 'STRATEGIC PLANNING': An assessment of new concepts and challenges," *Information Technology and Management*, 4 (2), 165–182.

Luftman, J. and T. Brier (1999), "Achieving and sustaining business–IT alignment," *California Management Review*, 42 (1), 109–122.

Luftman, J.N., P.R. Lewis, and S.H. Oldach (1993), "Transforming the enterprise: The alignment of business and information technology strategies," *IBM Systems Journal*, 32 (1).

Ma, C., D.C. Chou, and D.C. Yen (2000), "Data warehousing, technology assessment and management," *Industrial Management and Data Systems*, 100 (3), 125–134.

Madria, S., S. Bhowmick, and W.K. Ng (2003), *Web Data Management*. New York: Springer-Verlag.

Maes, R. (1999), *Reconsidering Information Management Through a Generic Framework*. Amsterdam: Universiteit van Amsterdam, Department of Accountancy and Information Management.

Maes, R. (2000), *Redefining Business: IT Alignment Through a Unified Framework*. Amsterdam: Universiteit van Amsterdam, Department of Accountancy and Information Management.

Manning, I.T. (1999), Data warehousing — Adopting an architectural view, and maximizing cost benefits, in *Data Warehousing*, SCN Education B.V. (Ed.). Germany: Vieweg.

March, S., A. Hevner, and S. Ram (2000), "Research commentary: An agenda for information technology research in heterogeneous and distributed environments," *Information Systems Research*, 11 (4), 327–341.

McFadden, F.R. (1996), "Data warehouse for EIS: Some issues and impacts," *29th Hawaii International Conference on System Sciences*, 2.

McLean, E.R. and J.V. Soden (1977), *Strategic Planning for MIS*. New York: John Wiley.

Meyer, D. and C. Cannon (1998), *Building a Better Data Warehouse*. Upper Saddle River, NJ: Prentice Hall PTR.

Mistry, H., P. Roy, S. Sudarshan, and K. Ramamritham (2001), "Materialized view selection and maintenance using multi-query optimization," in *Proceedings of the 2001 ACM SIGMOD International Conference on Management of Data*, New York: ACM Press, 30.

Mohania, M. and G. Dong (1996), "Algorithms for adapting materialized views in data warehouses," *Proceedings of the International Symposium on Cooperative Database Systems for Advanced Applications*, December.

Moody, D.L. and M.A.R. Kortink (2000), "From enterprise models to dimensional models: A methodology for data warehouse and data mart design," *Proceedings of the International Workshop on Design and Management of Data Warehouses*. Stockholm, Sweden.

Mooney, J.G., V. Gurbaxani, and K.L. Kraemer (1996), "A process oriented framework for assessing the business value of information technology," *Proceedings of the 16th Annual International Conference on Information Systems*; ACM SIGMIS Database, 27 (2), 68–81.

Mukhopadhyay, T., S. Rajiv, and K. Srinivasan (1997), "Information technology impact on process output and quality," *Management Science*, 43 (12), 1645–1659.

Muller, H.A., J.H. Jahnke, D.B. Smith, M.A. Storey, S.R. Tilley, and K. Wong (2000), "Reverse engineering: A roadmap," *Proceedings of the Conference on the Future of Software Engineering*. Limerick, Ireland, June 04–11, 47–60.

Murtaza, A.H. (1998), "A framework for developing enterprise data warehouses," *Information Systems Management*, 15 (4), 21–26.

Nah, F.F.H., X. Tan, and S.H. Teh (2004), "An empirical investigation on end-users' acceptance of enterprise systems," *Information Resources Management Journal*, 17 (3), 32–53.

Nemati, H.R., D.M. Steiger, L.S. Iyer, and R.T. Herschel (2002), "Knowledge warehouse: An architectural integration of knowledge management, decision support, artificial intelligence and data warehousing," *Decision Support Systems*, 33 (2), 143–161.

Niemi, T., L. Hirvonen, and K. Jaervelin (2003), "Multidimensional data model and query language for informetrics," *Journal of the American Society for Information Science and Technology*, 54 (10), 939–951.

O'Sullivan, O. (1996), "Data warehousing — without the warehouse," *ABA Banking Journal*, 88 (12).

Oakley, S. (1999), "Data mining, distributed networks, and the laboratory," *Health Management Technology*, 20 (5), 26.

Oates, J. (1998), "Evaluating data warehouse toolkits," *IEEE Software*, 15 (1), 52–54.

Papp, R. (2001), *Strategic Information Technology: Opportunities for Competitive Advantage*, Hershey, PA: Idea Group.

Peacock, P.R. (1998), "Data warehouses and marts," *Marketing Management*, 6 (4), 13.

Pedersen, T.B. and C.S. Jensen (1998), "Research issues in clinical data warehousing," *Scientific and Statistical Database Management*, 43–52.

Pliskin, N., C.T. Romm, A.S. Lee, and Y. Weber (1993), "Presumed versus actual organizational culture: Managerial implications for implementation of information systems," *The Computer Journal*, 36 (2), 143–152.

Poe, V., P. Klauer, and S. Brobst (1998), *Building a Data Warehouse for Decision Support*. Upper Saddle River, NJ: Prentice Hall.

Pollalis, Y.A. (2003), "Patterns of co-alignment in information-intensive organizations: Business performance through integration strategies," *International Journal of Information Management*, 23 (6), 469–492.

Porter, M.E. and V.E. Millar (1985), "How information gives you competitive advantage," *Harvard Business Review*, 63 (4), 149–160.

Powell, J.T.C. and A. Dent-Micallef (1997), "Information technology as competitive advantage: The role of human, business, and technology resources," *Strategic Management Journal*, 18 (5), 375–405.

Prahalad, C.K. and M.S. Krishnan (2002), "The dynamic synchronization of strategy and information technology," *MIT Sloan Management Review*, 43 (4), 24–33.

Praskey, S. (2001), "Service spotlight," *Canadian Insurance*, 106 (6), 26–29.

Rabinovich, E., J.P. Bailey, and C.R. Carter (2003), "A transaction-efficiency analysis of an Internet retailing supply chain in the music CD industry," *Decision Sciences*, 34 (1), 131–172.

Raghupathi, W. and J. Tan (2002), "Strategic IT applications in health care," *Communications of the ACM*, 45 (12), 56–61.

Rahm, E. and H.H. Do (2000), "Data cleaning: Problems and current approaches," *IEEE Data Engineering Bulletin*, 23 (4), 3–13.

Redman, T.C. (1995), "Improve data quality for competitive advantage," *Sloan Management Review*, 36 (2), 99–107.

Robertson, P. (1997), "Integrating legacy systems with modern corporate applications," *Communications of the ACM*, 40 (5), 39–46.

Roth, K., D.M. Schweiger, and A.J. Morrison (1991), "Global strategy implementation at the business unit level: Operational capabilities and administrative mechanisms," *Journal of International Business Studies*, 22 (3).

Roussopoulos, N. (1998), "Materialized views and data warehouses," in *ACM SIGMOD Record*. New York: ACM Press, 27.

Rundensteiner, E.A., A. Koeller, and X. Zhang (2000), "Maintaining data warehouses over changing information sources," *Communications of the ACM*, 43 (6), 57–62.

Ryan, S.D. and D.A. Harrison (2000), "Considering social subsystem costs and benefits in information technology investment decisions: A view from the field on anticipated payoffs," *Journal of Management Information Systems*, 16 (4), 11–40.

Rygielski, C., J.C. Wang, and D.C. Yen (2002), "Data mining techniques for customer relationship management," *Technology in Society*, 24 (4), 483–502.

Sammon, D. and P. Finnegan (2000), "The ten commandments of data warehousing," *The Database for Advances in Information Systems*, 31 (4), 82–91.

Samos, J., F. Saltor, J. Sistac, and A. Bardes (1998), "Database architecture for data warehousing: An evolutionary approach," *Database and Expert Systems Applications*, 1460, 746–756.

Sarkis, J. (2000), "An analysis of the operational efficiency of major airports in the United States," *Journal of Operations Management*, 18 (3), 335–351.

Schmidt, G. (2000), "Strategic, tactical and operational decisions in multi-national logistics networks: A review and discussion of modelling issues," *International Journal of Production Research*, 38 (7), 1501–1523.

Schniederjans, M.J. and J.L. Hamaker (2003), "A new strategic information technology investment model," *Management Decision*, 41 (1), 8–17.

Schubart, J.R. and J.S. Einbinder (2000), "Evaluation of a data warehouse in an academic health sciences center," *International Journal of Medical Informatics*, 60 (3), 319–333.

Segars, A.H. and V. Grover (1998), "Strategic information systems planning success: An investigation of the construct and its measurement," *MIS Quarterly*, 22 (2), 139–163.

Sen, A. (2004), "Metadata management: Past, present and future," *Decision Support Systems*, 37 (1), 151–173.

Sen, A. and V.S. Jacob (1998), "Industrial-strength data warehousing," *Communications of the ACM*, 41 (9), 28–31.

Sethi, V. and W.R. King (1994), "Development of measures to assess the extent to which an information technology application provides competitive advantage," *Management Science*, 40 (12), 1601–1627.

Shahzad, M.A. (1999), "Data warehousing with Oracle," in *Proceedings of SPIE — The International Society for Optical Engineering*, 3695.

Shankaranarayanan, G. and A. Even (2004), "Managing metadata in data warehouses: Pitfalls and possibilities," *Communications of the Association for Information Systems*. Atlanta, 14.

Shanks, G. and B. Corbitt (1999), "Understanding data quality: Social and cultural aspects," in *Proceedings of the 10th Australasian Conference on Information Systems*, B. Hope and P. Yoon (Eds.), Wellington, New Zealand: Victoria University of Wellington.

Shapiro, J.F. (2001), *Modeling the Supply Chain*. Pacific Grove, CA: Brooks/Cole-Thomson Learning.

Shi, D., Y. Lee, X. Duan, and Q.H. Wu (2001), "Power system data warehouses," *IEEE Computer Applications in Power*, 14 (3), 49–55.

Shim, J.P., M. Warkentin, J.F. Courtney, D.J. Power, R. Sharda, and C. Carlsson (2002), "Past, present, and future of decision support technology," *Decision Support Systems*, 33 (2), 111–126.

Shin, B. (2002), "A case of data warehousing project management," *Information and Management*, 39 (7), 581–592.

Shin, B. (2003), "An exploratory investigation of system success factors in data warehousing," *Journal of the Association for Information Systems*, 4 (170), 141–170.

Shin, N. (2001), "The impact of information technology on financial performance: The importance of strategic choice," *European Journal of Information Systems*, 10 (4), 227–236.

Sigal, M. (1998), "A common sense development strategy," *Communications of the ACM*, 41 (9), 42–43.

Sinn, W. (2003), "How to avoid 'bad data days'," *The CPA Journal*, 73 (7), 11.

Smaczny, T. (2001), "Is an alignment between business and information technology the appropriate paradigm to manage IT in today's organisations," *Management Decision*, 39 (10), 797–802.

Sperley, E. (1999), *Enterprise Data Warehouse*. Upper Saddle River, NJ: Prentice Hall.

Squire, C. (1995), "Data extraction and transformation for the data warehouse," in *ACM SIGMOD Record*. New York: ACM Press, 24.

Srivastava, J. and P.Y. Chen (1999), "Warehouse creation — A potential roadblock to data warehousing," *IEEE Publications on Knowledge and Data Engineering*, 11 (1), 118–126.

Srivastava, R.K., T.A. Shervani, and L. Fahey (1998), "Market-based assets and shareholder value: A framework for analysis," *Journal of Marketing*, 62 (1), 2–18.

Strong, D.M., Y.W. Lee, and R.Y. Wang (1997), "Data quality in context," *Communications of the ACM*, 40 (5), 103–110.

Sullivan, D. (2001), Document *Warehousing and Text Mining: Techniques for Improving Business Operations, Marketing and Sales*. New York: John Wiley.

Sumner, M. (2000), "Risk factors in enterprise-wide/ERP projects," *Journal of Information Technology*, 15 (4), 317–327.

Swift, R.S. (2000), *Accelerating Customer Relationships: Using CRM and Relationship Technologies*. Upper Saddle River, NJ: Prentice Hall.

Tallon, P.P., K.L. Kraemer, and V. Gurbaxani (2000), "Executives' perceptions of the business value of information technology: A process-oriented approach," *Journal of Management Information Systems*, 16 (4), 145–173.

Talluri, S. (2000), "An IT/IS acquisition and justification model for supply chain management," *International Journal of Physical Distribution and Logistics Management*, 30 (3), 221–237.

Tan, X., D.C. Yen, and X. Fang (2003), "Web warehousing: Web technology meets data warehousing," *Technology in Society*, 25 (1), 131–148.

Teo, T.S.H. and J.S.K. Ang (1999), "Critical success factors in the alignment of IS plans with business plans," *International Journal of Information Management*, 19 (2), 173–185.

Tillquist, J. (2002), "Strategic connectivity in extended enterprise networks," *Journal of Electronic Commerce Research*, 3 (2), 77–85.

Triantafillakis, A., P. Kanellis, and D. Martakos (2004), "Data warehousing interoperability for the extended enterprise," *Journal of Database Management*, 15 (3), 73.

Tryfona, N., F. Busborg, and J.G.B. Christiansen (1999), "starER: A conceptual model for data warehouse design," in *2nd ACM International Workshop on Data Warehousing and OLAP*. New York: ACM Press.

Tyagi, S. (2003), "Using data analytics for greater profits," *Journal of Business Strategy*, 24 (3), 12–14.

Umble, E.J., R.R. Haft, and M.M. Umble (2003), "Enterprise resource planning: Implementation procedures and critical success factors," *European Journal of Operational Research*, 146 (2), 241–257.

Van Den Hoven, J. (1998), "Data marts: Plan big, build small," *Information Systems Management*, 15 (1), 71–73.

Van Eck, P., H. Blanken, and R. Wieringa (2004), "Project GRAAL: Towards operational architecture alignment," *International Journal of Cooperative Information Systems*, 13 (3), 235–255.

Vander Vennet, R. (2002), "Cost and profit efficiency of financial conglomerates and universal banks in Europe," *Journal of Money, Credit and Banking*, 34 (1), 254–282.

Vassiliadis, P. (2000), "Gulliver in the land of data warehousing: Practical experiences and observations of a researcher," in *Proceedings of the International Workshop on Design and Management of Data Warehouses*, M. Jeusfeld, H. Shu, M. Staudt, and G. Vossen (Eds.). Stockholm, Sweden, June 5–6.

Vassiliadis, P., M. Bouzeghoub, and C. Quix (2000), "Towards quality-oriented data warehouse usage and evolution," *Information Systems*, 25 (2), 89–115.

Vassiliadis, P. and T. Sellis (1999), "A survey of logical models for OLAP databases," in *ACM SIGMOD*. New York: ACM Press, 28.

Vatanasombut, B. and P. Gray (1999), "Factors for success in data warehousing: What the literature tells us," *Journal of Data Warehousing*, 4 (3), 25–33.

Watson, H.J., J.G. Gerard, L.E. Gonzalez, M.E. Haywood, and D. Fenton (1999), "Data warehousing failures: Case studies and findings," *Journal of Data Warehousing*, 4 (1), 44–55.

Watson, H.J., D.L. Goodhue, and B.H. Wixom (2002), "The benefits of data warehousing: Why some organizations realize exceptional payoffs," *Information and Management*, 39 (6), 491–502.

Watson, H.J. and B.J. Haley (1998), "Managerial considerations," *Communications of the ACM,* 41 (9), 33.

Watson, H.J., C. Fuller, and T. Ariyachandra (2004), "Data warehouse governance: Best practices at Blue Cross and Blue Shield of North Carolina," *Decision Support Systems*, 38 (3), 435–450.

Weir, R., T. Peng, and J. Kerridge (2003), "Best practice for implementing a data warehouse: A review for strategic alignment," *5th International Workshop on the Design and Management of Data Warehouses*, Berlin.

Wells, D. and J. Thomann (1995), "The keys to the data warehouse," *American Programmer*, 8 (5), 9–17.

Wen, H.J., D.C. Chou, and D.C. Yen (1997), "Building a data warehouse: An overview," *Communications of the ICISA*, Fall, 25–35.

Westerman, P. (2001), *Data Warehousing: Using the Wal-Mart Model,* New York: Morgan Kaufmann. Widom, J. (1995), "Research problems in data warehousing," in *Proceedings of the Fourth International Conference on Information and Knowledge Management*. New York: ACM Press.

Williams, C. (1999), "Enhancing the information harvest — Data warehousing systems: How they work and what they can do for your company," *Bobbin*, Vol. 40.

Wixom, B.H. and H.J. Watson (2001), "An empirical investigation of the factors affecting data warehousing success," *MIS Quarterly*, 25 (1), 17–41.

Xu, Y., D.C. Yen, and B. Lin (2002), "Adopting customer relationship management technology," *Industrial Management and Data Systems*, 102 (8), 442–452.

Yang, J., K. Karlapalem, and Q. Li (1997), "Algorithms for materialized view design in data warehousing environment," *Proceedings of the 23rd VLDB Conference on Very Large Databases*. Athens, Greece.

Zaiane, O.R., M. Xin, and J. Han (1998), "Discovering Web access patterns and trends by applying OLAP and data mining technology on Web logs," *Research and Technology Advances in Digital Libraries*, 19–29.

Zeng, Y., R. Chiang, and D.C. Yen (2003a), "Enterprise integration with advanced information technologies: ERP and data warehousing," *Information Management and Computer Security*, 11 (3), 115–122.

Zeng, Y.E., H.J. Wen, and D.C. Yen (2003b), "Customer relationship management (CRM) in business-to-business (B2B) e-commerce," *Information Management and Computer Security*, 11 (1), 39–44.

Zhou, S., A. Zhou, X. Tao, and Y. Hu (2000), "Hierarchically distributed data warehouse," in *Proceedings of the Fourth International Conference on High-Performance Computing in the Asia-Pacific Region.* Washington, DC: IEEE Computer Society, 2.

Zorn, P., M. Emanoil, L. Marshall, and M. Panek (1999), "Mining meets the Web," *Online* 23 (5), 61–68.

Index

Milton Keynes UK
Ingram Content Group UK Ltd.
UKHW040059071024
449327UK00019B/661